SHIRLEY E. ANDERSEN

THE GREAT DEEP

Where Science and Scripture Collide

ISBN 978-1-961093-26-3 The Great Deep SC
ISBN 978-1-961093-27-0 The Great Deep e-book

Published by Silversmith Press—Houston, Texas
www.silversmithpress.com

SILVERSMITH
PRESS

CONTENTS

Acknowledgements...vii

Author's Note...ix

Introduction xi

Chapter One: A Circle Inscribed 1

    Who Inscribed a Circle On The Face of the Deep 1

    A Cleft Channel Encircling the Globe.. 9

Chapter Two: The Great Deep 11

    More Questions.. 13

    The Vaulted Dome.. 16

    The Earth Was Formless... 17

    For Truth is Truth 19

    The Great Void 22

    Four Rivers... 24

Chapter Three: The Illusive Garden of Eden.......................... 30

    The Great Deception.. 34

    The Father of Humanity 38

Names of the Pre-Flood Patriarchs:

Meaning & Significance 39

Memorialization of God's Promise.. 41

Chapter Four: Created in God's Image 44

A Corruption of Knowledge... 53

Chapter Five: Cain Kills His Brother 57

Cain's Ancient City of Erech... 61

Eve's Third Son, Seth 64

Righteous Enoch 68

Zion & The Pillar of Enoch 72

Chapter Six: Nephilim Were on the Earth in Those Days......... 76

"But Noah was a Righteous Man".. 80

The Stone Things... 86

Chapter Seven: Dynamic Synergy and Chain Reactions 91

Explosions from the Great Deep.. 99

The Illusive Anomaly—A Shift of Earth's Polar Axis 102

A Spreading Mid-Ocean Ridge or a Cleft Channel?.................. 107

Chapter Eight: The Mountain Tops Appeared 113

Excavated Ocean Basins and the Continental Boundary 116

Fossils and Fossil Fuels... 120

Chapter Nine: New Beginnings 131

Nimrod and the Tower of Babel 136

A Great Migration 142

CONTENTS

Chapter Ten: Oceanic Navigation Prior to the Ice Age........... 145

    Petroglyphs & Solar Flares 149

    A Sudden Freeze and Wooly Mammoths.............................. 152

    Who Survived the Ice Age.. 154

Chapter Eleven: Shattering Significance 160

    Abraham 165

Summary.. 171

ACKNOWLEDGEMENTS

Special thanks to my dear friend, Tammi and her husband Mika, for their encouragement and financial support that made it possible to publish this book. They had faith in me when I didn't believe in myself and provided a place of refuge at their beautiful B&B called Peace of Maui. This location became a haven for devoting my time and attention to attaining this goal. Please check out their website at www.peace ofmaui.com.

I am also grateful for Tomas Zeman who expanded my understanding about the literal accuracy of God's word resulting in a 1995 publication of the book we wrote together named "Genesis Geology." This was the beginning of my research and exploration of human history using Holy Scripture.

I also want to thank Steve Dorris, Shena Koch, and Pastor Mac of Calvary Chapel Kaneohe for taking time to review and comment on my first draft. Your suggestions and feedback gave me the courage I needed to see this project through to completion.

My heart-felt appreciation extends to Ellie, the owner of Tutu's Pantry who is an uplifting, supportive soul who provided a roof over my head in times of need, along with several other beloved friends like Ruth Kuni.

Most of all, I am grateful for the kind heart and sharp minds of my four sons, Raf, Quincy, Jason, and Skyler. You are a precious gift.

And thank you, Jesus, for the answered prayers and for desiring to share Your throne with us, "Behold, I stand at the door and knock! If anyone hears my voice and opens the door, indeed I will come into him and dine with him, and he with me. The one who conquers, I will grant to him to sit down with me on my throne, as I also have conquered and have sat down with my Father on His throne" (Revelation 3:20-21).

AUTHOR'S NOTE

Throughout my journey of research, I gleaned information from non-canonical books like the *Book of Enoch*, the *Book of Jasher*, *Antiquities of the Jews*, and *The Cave of Treasures*, to name a few. These ancient scribes contributed their perspectives and experiences in a way I found useful for a fuller comprehension and understanding of Holy Scripture as a historic document. My research was enhanced by a book called *The Old Testament Pseudepigrapha* by James H. Charlesworth, who elaborates upon the value of these writings, *"Mere perusal of the biblical books discloses that their authors depended upon sources that are no longer extant [. . .] after 90 A.D. there were still debates regarding the canonicity of such writings as the Song of Songs, Ecclesiastes, and Esther, but it is not clear what were the full ramifications of these debates. It seems to follow, therefore, both that the early pseudepigrapha were composed during a period in which the limits of the canon apparently remained fluid at least to some Jews, and that some Jews and Christians inherited and passed on these documents as inspired. They did not necessarily regard them as apocry-phal, or outside a canon.*"[1]

---

[1]James H. Charlesworth, *The Old Testament Pseudepigrapha*, xxiii.

INTRODUCTION

My father proclaimed to be an atheist, and he used to say, "when you're dead, you're dead, and that's it; life is over." He also told me the Bible was a story most likely made up by people who needed something to believe in, "you know . . . so they can get through troubled times." Sometimes I look back and wonder if he was just playing the devil's advocate because my father was such a generous, caring, and helpful person who lived through many hardships. I can only hope that before his last breath, he chose to receive forgiveness and was blessed with the gift of eternal life.

My mom was raised a Christian. I remember her tiny child-hood Bible resting upon the shelf of a bar my dad had built for entertaining friends. About the age of seven, I was very curious about this little book. I remember being captivated by the artistic illustrations. The cover of her Bible displayed an image of baby Moses being lifted out of the water by a servant of the pharaoh's daughter.

Religion and the Bible did not have much of an influence upon my life growing up, but now I look back and can see that from an early age, God definitely had a plan for my life. The significance of Moses' life was recently revealed to me when I discovered God's mighty plans were accomplished through him in spite of his stammering and stuttering. This was when I realized God could even use me to be a spokesman for Him. I am not eloquent of speech; I am not a college graduate; I don't know much about science, geology, and physics; but Almighty God, the Creator of heavens and the Earth, the King of Kings, laid it upon my heart to share with you the truth about the accuracy of His Word.

I believe the story of Moses reveals God wants His voice to be heard through us in spite of our imperfections. God's voice is heard through our thoughts, through the words we speak or hear being spoken, even words spoken within our mind. As we desire alignment with His will and pray for His will being done here on Earth, as it is in heaven . . . the manifestation of the kingdom of righteousness will be birthed. Utterance of His Word will not return void. I now understand my childhood fascination with the picture on the cover of my mother's Bible and believe it was a foreshadowing of things to come because I would be called upon to announce our Heavenly Father's code of perfection.

By the age of seventeen, I was bored silly with high school and dropped out just prior to graduation. As I look back upon my life, I have no regrets. Deep in my heart, I knew there had to be something more to life than what my schooling had taught and conditioned for me to become . . . a go-for-this or gopher-that secretary. I was so disappointed with the outcome of my education. I enrolled in a few college evening classes for business management, but they didn't pique my interest either. I felt robbed by the system of education and decided to fly from Ontario, Canada to Hawaii expecting to find dirt roads and grass shacks. I am so grateful for a close friend who guided me to land on Maui. I was surprised to discover it was far more civilized than I had imagined, but at this point in time, there were only two traffic lights on the whole island ! It was the year of 1975, and this paradisiacal place was where I became exposed to knowledge from teachers who had traveled the globe gathering wisdom from various cultures.

Soon after arriving here, I learned to read Tarot Cards, became an astrologer's apprentice, studied the healing power of sound with cymatics, and learned how to combine medicinal herbs. I gathered an understanding about the ley lines of Planet Earth, along with knowledge of energy flowing through meridians in the human body. I was introduced to Kirlian photography and had fun capturing the auras of people as their radiant energy was exposed to film and revealed with photographs. One of the special highlights of my life took place in 1976 when I brought my Kirlian aura photography unit to the Honolulu Renaissance Fair at Diamond Head Crater. At about the same time I was also gifted a Radionics machine. My father was convinced the electromagnetic waves generated by this device were extremely helpful for managing his pain and stress.

I was passionate about the healing arts and studied to become a licensed massage therapist. As I learned about human anatomy, I found it disturbing that this valuable information had been omitted from my education. I questioned why there had been no mention whatsoever about the process of nutrients being absorbed at a cellular level. I still believe our children today would live much healthier lives if this knowledge was shared with them in a simple way at an early age. As I found my way through life, and the depth of my knowledge expanded, one question persisted. . . . Who is God?

About seven years after I arrived on Maui, a life-transforming event took place while I was reading a *National Geographic* magazine article published in 1980. It may sound strange, but God touched my heart in the most profound and intimate way as I read this fascinating story about the Shroud of Turin. The Shroud is believed by millions to be the sacred cloth that once enveloped Christ's body as Joseph

My Aura

of Arimathea placed it within the tomb. This *National Geographic* article was presenting the different views of historians and scientists from around the world, who were investigating and commenting on how the faint outline of a tortured man could have been left as an imprint upon the cloth covering his dead body. As they tried to determine if it was authentic, most agreed it did not show evidence of tampering or the addition of artificial pigments.

Current pictures of the Shroud reveal intricate details about the outline of Christ's physical form. Some scientists examining this cloth, known as "The Shroud of Turin," concluded there was "stronger evidence of a faint scorch, but how it could be produced has not been determined." It was as if Christ's image had been burnt into the sacred cloth enveloping His entire body. There were blood stains where nails pierced His wrists, and evidence of blood dripping from His forehead where a thorny crown was pressed into His skull, stains where fluids poured out from His side after His body was pierced by the spear of a soldier, "but one of the soldiers pierced His side with a spear, and immediately there came out blood and water" (John 19:34).

As I read this *National Geographic* article about the shroud and this information became absorbed, a profound revelation took place. Within a fraction of a second, I perceived electrons orbiting atoms within the nucleus of every cell in the body at nearly the speed of light—billions of whirls in a millionth of a second. I thought about the radiance of a new couple in love, the glow of a pregnant woman, and suddenly realized these phenomena were just a shadow of the greatness of God's love for His Son and all of humanity. This was when I realized the synergistic power of God's love had transformed Christ's body into a being of light—that God's gift of eternal life had been made known through His only begotten Son, whose imprint was burned into the Shroud of Turin. This sacred cloth revealed to me that somehow Christ's body transformed into a being of light. It was a life-changing experience. I fell to my knees weeping and cried out to Him, *"Please Lord, forgive me! I am so extremely sorry for ever having doubted You."*

In 1898 a photographer, while developing pictures of the Shroud in his darkroom, discovered the cloth actually contained a negative image.

> *Secundo Pia examined his first glass plate negative as it emerged from the developing bath; he almost dropped it in shocked excitement. He was looking not at the usually unrealistic, confusing, photographic negative, but at a clear positive image . . . a negative image? Hundreds of years before the invention of photography? The idea that the shroud was a hoax suddenly seemed less plausible, for how could a medieval artist have produced a negative image, and why would he choose to do so?* (Kenneth F. Weaver, "The Mystery of the Shroud," *National Geographic* 157, [June 1980]: 743)[1]

---

[1] Kenneth F. Weaver, "The Mystery of the Shroud," *National Geographic* 157, [June 1980]: 743

The presence of the Holy Spirit alighted upon my soul as I comprehended and accepted Jesus truly was the Lamb of God who had sacrificed His life to pay the price for my sin (John 1:36). I believe the Shroud of Turin was preserved as a testament to the truth of Christ's resurrection. The Shroud currently resides within the Cathedral of Saint John the Baptist in Turin, Italy. I find it truly fascinating God spoke to me and touched my heart through research and understanding of auras. It was confirmation God Almighty had taken the time to intimately know who I was, to accept and love me for who I am.

It was at this point in time the Bible was no longer a fairytale. This was the beginning of my walk with Christ and searching to know Him. A more recent *National Geographic* article about the Shroud of Turin states:

> *Over the 117 years since a photographic negative of the linen unexpectedly revealed the image of a tortured body, ranks of physicists and chemists have weighed in on the fabric's age and the image's composition. Forensic pathologists, microbiologists, and botanists have analyzed its bloodstains, along with specks of dirt and pollen on its surface. Statisticians have combed through mountains of data.*[1]

---

[1] Frank Viviano, "Why Shroud of Turin's Secrets Continue to Elude Science," *National Geographic*, (April 2015): https://www.nationalgeographic.com/history/article/150417-shroud-turin-relics-jesus-catholic-church-religion-science last accessed 11-17-2022.

The article goes on to conclude that so far, researchers have been unable to dismiss the shroud as a forgery.[1]

After having this experience in 1983, I joined the Upcountry Christian Fellowship in Pukalani. I was hungry for God's Word. I became enamored with radio shows like those of Derek Prince, but it took some time to figure out what I should do with all of the books I had purchased to calculate astrology charts. A close friend encouraged me to seek wisdom and guidance from God, who had created the stars (Psalm 147:4), rather than trying to guide my or anyone else's life by the stars created and named by Him. I took his advice to heart, but these astrology books were worth a lot of money. I felt perplexed and finally looked up to the heavens, asking God what I should do. To my great surprise, there was an immediate response that totally grabbed my attention as I heard His thunderous voice from above. It was loud and clear as if He were right there in the room with me saying, "Believe in me, and I will show you the way. Trust in Me for I have more wisdom and knowledge for you than you could even fathom." After this experience I had no doubt or any reservations about what needed to be done. These astrological books for "divining" became worthless. I got rid of them and never looked back. "Let now the astrologers, those who prophecy by the stars, those who predict by the new moons, stand up and save you from what will come upon you" (Isaiah 47:13).

---

[1] https://commons.wikimedia.org/wiki/File:Shroudofturin.jpg Public domain author's life plus 100-years or fewer

To this day God has not let me down with His promise of more wisdom and knowledge than I could imagine. As I began questioning and thinking . . . "if God is for real, His Word as preserved within Holy Scripture must be accurate, true, and in alignment with what we observe of His creation . . . If the Bible was not myth, there could be no discrepancy." As this mindset persisted, I was led on an amazing expedition of discovery as layer upon layer of evidence for the "literal" accuracy of His Word was revealed.

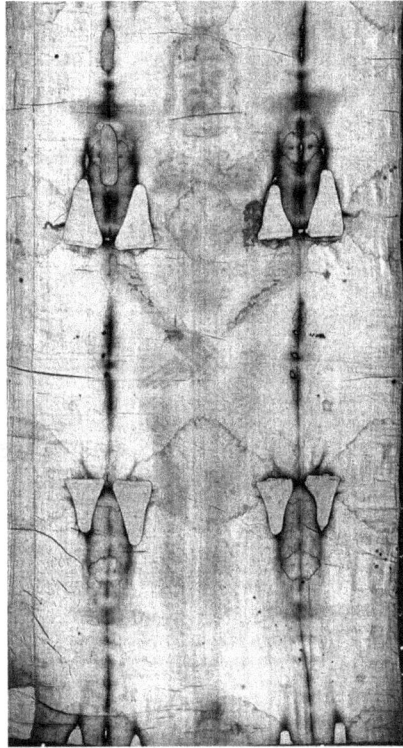

Shroud of Turin

Descriptive details about the ocean floor, insight to the dynamics of solar flares, an understanding of how every single mineral and element known to man became suspended within ocean water; all of this and more was brought to light as His Word became a siphon for investigating a complex array of information.

The purpose for my writing this book is to encourage Christians to be secure in their faith and know God more intimately. I invite you to an exploration of our world and human history from a completely new and groundbreaking perspective through the lens of Holy Scripture. I am convinced the information I have to share is biblically based and astonishingly reveals how God's Word accurately aligns with what we can observe of His creation. My hope and prayer is that my endeavor leads people to see that from the beginning of creation, God Almighty has revealed His plan for the redemption of souls.

It is my hope and prayer that contemplation of this alternative perspective, and seeing things from a different viewpoint, will make it increasingly easy to rest assured and know Almighty God has spelled out everything from beginning to end. He is the author and finisher of our faith and laid out His master plan for all to see. I am convinced the Bible is a historic record and not just something made up by people trying to cope with difficult times. Prepare to take a journey from the beginning of creation to our present day as we begin untying the knots and mysteries of a great deception.

My story reveals the truth about the root of who we are as brothers and sisters of a noble human family. My prayer is for unity and acceptance of God's only begotten Son who sacrificed His life for the redemption of our souls. We will travel back in time to when "the great deception" began. It is a deeply rooted story. One prevailing for thousands upon thousands of years, and today, just as in the days of Noah, we find ourselves living during a time of spurious forces with a global influence, and God is making the last call for people to wake up and board His ark of salvation. "In the last days mockers will come with their mocking . . . saying, 'Where is the promise of His coming?' . . . it escapes their notice that by the word of God the heavens existed long ago and the earth was formed out of water and by water" (2Peter 3:3-5).

"A revolution is coming:
A revolution which will be peaceful, if we are wise enough,
compassionate if we care enough,
successful if we are fortunate enough—but a revolution which is coming whether we will it or not.
We can affect its character; we cannot alter its inevitability."
Robert F Kennedy—1964
Speech at the University of Pennsylvania

What you are about to read is a new way of looking at geology, science, and human history. It opposes popular opinion and based upon the literal accuracy of Holy Scripture.

CHAPTER 1
A CIRCLE INSCRIBED

He Inscribed a Circle On the Face of the Deep.
(Proverbs 8:27)

The Earth presents itself as a riddle, and my story begins with a map of the ocean floor. One created under the lead of geophysicist Maurice Ewing. Mapping of the ocean floor began with artist/cartographer Marie Tharp and geologist/cartographer Bruce C. Heezen. They paired up to systematically map the entire foundation of Earth at the ocean floor way back in the '60s. Their preliminary research and drawings began in 1957. This project was backed by the United States Naval Office of Research. What Heezen and Tharp discovered at the ocean floor was a cleft channel encircling the entire globe. Their discovery was monumental, but the map was hidden from public view for quite some time, supposedly due to the cold war with Russia and secret intelligence regarding submarine navigation. But I believe if people had become aware of this magnificent feature at the ocean floor prior to science explaining it away with voraciously promoted hypotheses about our planet evolving over billions of years, the profound significance of Proverbs 8:27 and chapter 38 of Job may have been made known.

"Soon Tharp was puzzling over the same kind of strange mountain ranges—cleft in the middle—that the British scientists on the John Murray Expedition had noticed. In places this cleft was as deep as the Grand Canyon. . . . They realized they were looking at something really big: in fact, the sea bottom revealed how the planet was formed."[1] A map of the North Atlantic Ocean was completed by Tharp in 1957, followed by maps of the South Atlantic and Indian Ocean by 1964. "Marie Tharp, as it turns out, is the one who really turned oceanography on its head. She mapped the seafloors and made the key discovery that led others to prove Wegener's drift theory."[2] The "drift" theory proposed by Alfred Wegener was that the Earth was once a supercontinent called Pangaea. The word Pangaea is an ancient Greek term meaning all land. Wegener proposed the supercontinent known as Pangaea was broken apart long ago and continents moved, horizontally, across the face of Earth to their current positions. His hypothesis was criticized and not widely accepted until the late '60s.

Heezen and Ewing were somehow enticed to play a key role "in proving Wegener's continental drift idea was

---

[1] Val Ross, *The Road to There: Mapmakers and Their Stories*, (Tundra Books).

[2] Ron Miksha, "The Death of Heezen," *The Mountain Mystery*, (June 2014): https://mountainmystery.com/2014/06/21/the-death-of-heezen/.

right, yet both remained skeptical about the whole scheme. Heezen believed the Earth was expanding; Ewing wasn't sure what to believe and likely doubted continents move at all."[1] What we currently observe of Earth's geologic features at the ocean floor does not give credence to Wegener's logic-based ideas, and yet his hypothesis became widely accepted when combined with Tharp's elaborately drawn maps of the ocean floor. It was at this point in time when Wegener's idea of continental drift became known as a new discovery called plate tectonics. The revealing of a mid-ocean rift with right angled fracture zones led many scientists "to accept the once violently controversial Wegener theory of continental drift."[2] Plate tectonics would become the consummate scientific discovery of our age. It exploded into mainstream media with an introduction by *National Geographic* in 1967, which highlighted Tharp and Heezen's mapping of the ocean floor. Plate tectonics was a theory that lent credence to the idea of Earth being millions, perhaps even billions, of years old.

The plate tectonics hypothesis (a.k.a. sea floor spreading) asserts the Mid-Ocean Rift is a place where the seafloor is spread apart by upwelling magma. It is proposed that over millions of years, new ocean floor is being produced and

---

[1]Ron Miksha, "The Death of Heezen," *The Mountain Mystery*, (June 2014): https://mountainmystery.com/2014/06/21/the-death-of -heezen/.

[2]Samuel W. Matthews, "Indian Ocean Unveiled On a Dramatic New Map—Science Explores the Monsoon Sea," *National Geographic* 132, no. 4: 567.

destroyed by a circulating system of convection cells. And, because the Earth is not growing any larger as the sea-floor spreads apart, it is proposed older sections of it are subducting back into the Earth at the edges of continents. One of the ways they substantiate this concept is with images showing the northwest coastline of Africa fitting like the piece of a puzzle into the outline of a shape formed by the Gulf of Mexico. With one quick glance at a map of the world, it is easy to see how the shape of these two continents would fit together and think this theory makes sense. But if you examine all of the other continents of our planet, you will see that none of them fit together like this. Upon closer investigation, more discrepancies arose and became evident with the plate tectonics hypothesis:

1) Molten magma is not currently observed to be upwelling in the cleft channel at the ocean floor and pushing continents apart, as is proposed by the theory. For example, magma upwelling within the cleft channel at the ocean floor has created a volcanic island above sea level we know as Iceland. Oh and by the way, doesn't it seem illogical to expect that magma upwelling in the rift valley at the ocean floor would leave behind a grand canyon encircling the globe?

2) The towering pillars adjacent to this encircling canyon are made of basalt, which starts out as molten lava. Until it cools, molten lava is in a thick, fluid-like state, comparable to cake batter. So if the molten lava was not poured or injected into a mold, like batter poured into a cake

pan, then there's no way it will form pillars. Just like cake batter poured onto a flat surface is unable to form a cake with tall sides. For molten lava to have formed towering pillars, it definitely would need to have been injected within a mold at some point in time. The concept of molten magma upwelling at the Mid-Ocean Rift, pushing continents apart . . . leaving a gaping canyon behind, and forming towering pillars just doesn't make sense.

3) All continental land above sea level is contained within an unbroken boundary known as the continental margin.

We will explore these challenges in detail, but in the beginning, when God's Word seemed so accurate and true, I was left questioning how a theory like sea floor spreading could be taught in schools as fact and why wasn't there anyone questioning it? Or perhaps there were people who disagreed with this hypothesis, but their perspectives and opinions were not widely published and promoted. For example, geologist Lester King asserts that "from the interior of a sphere, there is only one way out: it is up towards the surface. Hence the active tectonics of the Earth are expressed radially or at the surface, vertically."[1]

Tharp's work attracted the attention of the National Geographic Society, which wanted to include a supplemental map of the Indian Ocean in their October 1967 issue. This

---

[1]Lester C. King, *Wandering Continents and Spreading Sea Floors on an Expanding Earth*, (New York: John Wiley & Sons Ltd., 1983), 16.

revolutionary article promoted the theory of plate tectonics as fact and says, "Here on the thin sea bottom, geophysicists say, Earth's crust is being pulled apart and new rock added, welling up in molten form from the underlying mantle of the planet," and "the Mid-Oceanic Ridge is now regarded as the longest continuous feature of the earth's solid face."[1] I found the timing for the creation of this new theory called plate tectonics rather dubious, as it coincided with *National Geographic*'s publication about Tharp and Heezen's mapping of the ocean floor. It was in 1967 plate tectonics gained so much acceptance that it revolutionized the way our planet was perceived by science and the world at large. "The Plate Tectonics Revolution was the scientific and cultural change which developed from the acceptance of the plate tectonics theory. The event was a paradigm shift and scientific revolution."[2] Question is, was the promotion of this map by *National Geographic* merely coincidence or was it contrivance? For who would dare question a scientific hypothesis promoted by mainstream media like *National Geographic* at the time, one that was endorsed by scientific authorities?

I remember my partner Tomas being met with opposition when questioning his college geology professor about the theory of plate tectonics. Perhaps like him, you too have

---

[1]Samuel W. Matthews, "Indian Ocean Unveiled On a Dramatic New Map—Science Explores the Monsoon Sea," *National Geographic* 132, no. 4: 564.

[2]Arturo Casadevall & Ferric C. Fang, "Revolutionary Science," *mBio* 7, no. 2: doi:10.1128/mBio.00158-16.

experienced a feeling of humility after asking questions you thought were intelligent, but the person responding had an attitude of superiority. One that often plays out like, "I know more than you do because I have a science degree." These are the exact words used on the show called "Ask Dr. Science . . . He knows more than you do!" This television show was brilliant and hilarious. Dr. Science says it best, "There is a thin line between ignorance and arrogance," and he assures us he has managed to erase that line. One of my favorite episodes can be found on YouTube; it is called "Ask Dr. Science: National Science Test 1987" by Duck's Breath Mystery Theater. Warning, you may end up rolling on the floor with laughter! But this truly is no laughing matter because the process of teaching a set of beliefs without any acceptance of criticism, then belittling those who question what is being taught, is typically understood and classified as brainwashing.

It appears to me as though Tharp and Heezen's enormous discovery was being used to promote the notion our planet had evolved over millions, perhaps even billions, of years. I remember the day Tomas picked up the Bible; he shook it, and said, "There must be something in here that explains it," and sure enough, he was right. After many prayers, discussions, and searching the Bible for answers, God began to reveal His creation with the prophetic words of Job 38:4, "Where were you when I laid the foundation of the earth? Tell Me, if you have understanding. Who set its measurements, since you know? Or who stretched a line on it?" This Bible verse piqued our attention, for we could plainly see the floor of the ocean's map revealed the foundation of Earth with a line

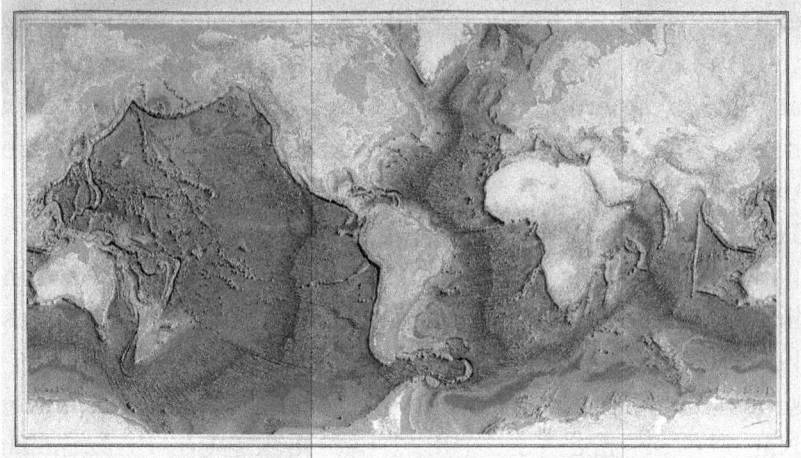

Map of the Ocean floor

carved through the center of the Atlantic Ocean. We fervently searched the Bible for more clues and found Proverbs 8:27, "When He established the heavens I was there, When He inscribed a circle on the face of the deep."

Today this circle inscribed upon the face of the deep is in plain sight for all to see, not only with the floor of the oceans map created by Tharp and Heezen but now we also have satellite images from outer space revealing the ocean floor. There is a line through the middle of the Atlantic Ocean stretching south and wrapping around the Horn of Africa before it heads north, then just south of the Red Sea it pitchforks to form an upside-down Y. The north stem of the upside-down Y enters the Red Sea and meets up with the Great Rift Valley of Africa. The eastern branch to the right of the upside-down Y loops beneath India and Australia, and then it continues to encircle the globe as it traverses the Pacific Ocean and merges with the North American Cordillera, adjacent to Baja California (see map).

A Cleft Channel Encircling the Globe

Now the very word "inscribe" means to write or carve into something so as to create a permanent record. I propose for your consideration that the "Floor of the Ocean" map and satellite images of Earth reveal God's carved inscription within the very foundation of Earth at the ocean floor. Maurice Ewing describes this incredible feature as "a cleft channel encircling the globe like the seam on a baseball."

Now let me ask you to consider if it is even possible for molten magma flowing at Earth's surface to push continents apart and form towering pillars at the ocean floor? Or is it more likely we have been influenced to accept scientific "theories" as fact, without being encouraged to use observation, logic, and critical thinking skills to question what we are told? Why is it that details recorded in Holy Scripture about how the Earth was formed are not taken into consideration? Could it be we are so enmeshed within a "cancel culture" working toward negating the Word of God that we do not recognize the literal accuracy of the Bible?

When I first accepted Christ as my Savior, I had reservations about some of the stories in the Bible being myth. I also gave some thought as to how different languages over thousands of years may have resulted in translation errors or the stories being misconstrued. But now I had a map and confirmation the Word of God was more literal, accurate, and true than what I first believed.

Sharing these concepts and discoveries was a call seemingly impossible to achieve, and beyond my capability. Obstacles in my personal life appeared insurmountable. There was violence, upheaval, and a multiplicity of distractions, but I could not abandon what God had laid upon my heart and kept nudging me to do. My persistence came from believing this mission could be accomplished with God's sovereignty.

Eventually, with the help of friends, I heard Him say, "Are you willing to complete this work?" I raised my hand and said, "Yes Lord, send me." I volunteered despite my inadequacies and imperfections, and God provided a way. I believed and trusted Him to take my fallen nature, redeem and restore it for His purposes. Then, many biblical verses encouraged my heart as prophetic revelations seemed to surface with persistent exploration of His Word.

CHAPTER 2

THE GREAT DEEP

The entire ocean floor is composed of towering pillars of basalt with bases sunk deep into the foundation of Earth. Job 38:6 prompts us to ask this question about His creation of the foundation, "On what were its bases sunk? Or who laid its cornerstone?" A cornerstone represents a right angle, and we can see the ocean floor pillars are at right angles to the Mid-Ocean cleft channel. Don't you think it is fascinating to observe this geologic feature at the ocean floor as it is described with the word "cornerstone" in Holy Scripture? These right-angled constructs at the ocean floor are classified and categorized by science as "fracture" zones: "The Mid-Ocean Ridge is cut by deep right-angle slashes which extend along the sea floor on both sides of the ridge. These are called fracture zones"[1]

Back in 1977 Bruce Heezen died as he explored the face of the deep in a submarine near Iceland. That same year Marie

---

[1]Hershell Nixon & Joan Lowery Nixon, *Land Under the Sea*, (New York: Dodd, Mead & Co., 1985), 43.

Tharp published a comprehensive map of the foundation of Earth called "The World Ocean Floor Map." Heezen's death at fifty-three years of age was untimely and surrounded by unusual circumstances. There had been feuds between Heezen and Maurice Ewing, the director of research on this project. As a result, Ewing cut off funding and marine data from Heezen. Quite often I wondered if at some point in time Heezen understood what Holy Scripture had to say about the foundation of Earth. My suspicions grew after discovering he had written a book called *The Face of The Deep*. For these are the exact words used in Proverbs 8:27 to describe the ocean floor, "When He established the heavens, I was there, When He inscribed a circle on the face of the deep."

Heezen's naming of his book may be coincidental, but my stumbling upon it twenty years after starting my investigation was a total game changer. By this time, Tomas and I were divorced. Our children were grown, independent adults. I really didn't have much hope of our book having any impact or significance. But every now and then, I would get inspired and share this information with friends. I just could not stop thinking that someday God would provide a way for it to be published. Why would He have revealed these things to me and placed such inspiration and passion within my heart? And now I happened to be holding in my hands a book written by Bruce Heezen called *The Face of The Deep*! I couldn't stop thinking and feeling as though God had called me to carry out this mission.

So in 2013, I began rewriting and creating artistic illustrations to reveal what I saw being described by Holy Scripture.

The realization that Tharp and Heezen's discovery had been articulately described by God's prophetic Word was monumental. As I continued researching and writing, I became convinced God was revealing His inscription upon the face of the deep during these last days to wake up a multitude of souls to the truth about who He is.

More Questions

It made sense that towering pillars at the ocean floor must have been formed by injection, but one question remained, what had provided a mold? We understood a total transformation of Earth had taken place during Noah's flood, but it took months of deliberation, prayer, and searching Scripture before there was a striking realization. Our minds lit up as we comprehended the significance of Psalm 18:15 along with 2 Samuel 22:16, "Then the channels of water appeared, And the foundations of the world were laid bare."

While pondering this verse, it became evident there were no ocean basins prior to the explosion of all the fountains of the great deep, and the foundation of Earth became exposed when the channels of water appeared, as they exploded from the great, deep, underground chambers (Genesis 7:11). A rush of excitement coursed through my veins as the weight of this discovery settled, and I began to comprehend how God's Word outlined a precise description of Earth's creation as God had designed our planet for such a transformation as this (Genesis 7:11). I recall a feeling of being set free from ambiguity. My apprehension about completely trusting in the literal accuracy of God's Word melted.

I stood at attention as His words presented articulate descriptions of Earth's geologic features. Psalm 18:15 is confirming our planet was "All Land." This verse of the Bible suggests the foundation of Earth was covered by an overlying crust, and it was this overlying crust that provided a mold for the injection of molten lava to form pillars of basalt, those we can now see exposed at the ocean floor. Psalm 18 explains Earth's pillar-like foundation became exposed when all the fountains of the great deep burst forth.

> *And the pillars of the earth were firmly established* (Job 38:4); *Tremble before Him, all the earth; Indeed, the world is firmly established, it will not be moved* (1 Chronicles 16:30); *THE LORD reigns. He is clothed with majesty; The LORD has clothed and girded Himself with strength; Indeed, the world is firmly established, it will not be moved.* (Psalm 93:1); and Psalm 96:10, *Say among the nations, "The LORD reigns." The world is firmly established, it cannot be moved; He shall judge the people righteously.*

Over time, all of our questions were answered with the help of Holy Scripture. Inspired by the Word of God, I began to draw artistic illustrations as I visualized the dynamics of Earth's formation. Chills ran up and down my spine as the enormity of this discovery became clearer with every prayer and investigation of Holy Scripture for evidence of the truth and accuracy of His Word. The excitement was overwhelming. At times it felt like we should be standing on rooftops and shouting out about the truth and accuracy of God's Word!

Looking back, the progression of how we understood His Word was logical. Every piece of the puzzle fit precisely into place and gave a clue about what to look for with the next piece. Another huge realization taking place in the very beginning of our research was the significance and meaning of the word "void" in Genesis 1:2. What we discovered is this verse about Earth being void was not describing a chaotic state but rather a cavity or recess of interconnected, empty (void) chambers waiting to be filled with water from the surface on day three of creation. "In the beginning God created the heavens and the earth, And the earth was formless and void" (Gen 1:1-2a).

It all began to make perfect sense! Wouldn't God Almighty have created our planet in such a way so as to accommodate His unfolding plan? "For thus says the Lord, who created the heavens (He is the God who formed the earth and made it, He established it and did not create it in vain)" (Isaiah 45:18). A working model of pre-flood Earth must have had a place for all the water to go on day three of creation when the dry land appeared. "He gathers the waters of the sea together as a heap; He lays up the deeps in storehouses" (Psalm 33:7). Jesus mentions Moses as the author of the book of Genesis and tells us if you don't believe in the Genesis account of creation written by Moses, you won't believe in Me. "For if you believed Moses, you would believe Me; for he wrote of Me.

But if you do not believe his writings, how will you believe My words" (John 5:46-47).

The Vaulted Dome

Mystery after mystery began to unravel as I saw processes about the creation of Earth unfolding like a movie in my mind. The intense pressure of steam spiraling—swirling—forcing its way through cavernous channels resulting in the formation of a great labyrinth of interconnected chambers beneath the "formless" surface of Earth. Gaseous explosions within these chambers would have created pockets of caves as the molten foundation formed. Eruptions of mineral-rich steam escaping through vents at Earth's surface blasted upward to hit a freezing cold wall of outer space. It is conceivable, with the total absence of sunlight on the second day of creation, that steam rapidly rose about two miles up from Earth's surface before colliding with the ice-cold pressure of outer space. This is where the mineral-rich steam compressed into a translucent sphere. What is described by Amos 9:6 as "the vaulted dome." Holy Scripture refers to this protective canopy as the firmament, and it is mentioned seven times in the Bible:

> *So God made the vaulted dome, and He caused a separation between the waters which were under the vaulted dome and between the waters which were over the vaulted dome, and it was so—and God called the vaulted dome "heaven." And*

there was evening, and there was morning, a second day.
(Gen 1:7-8)

Does not the very word "vault" bring to mind the concept of metal? The superconductive metallic qualities of this ice canopy would have facilitated a free flow of electrons providing for its suspension above the dipole magnetic field of Earth's north and south poles.

The Earth Was Formless

Not only was there a void resulting from processes creating our planet, but Scripture also articulates that in the beginning, the Earth was created as a formless sphere. Not in the sense that "formless" was a meaningless or chaotic state, but rather our planet was created without form. Its surface was like a smooth, round ball. There weren't any of the tremendous geologic variations we observe today. This is because vertical uplift of the mountain ranges and sinking of the abyssal trenches had not yet taken place. As described by Psalm 104:8 it was during the great flood Event when "the mountains rose, and the valleys sank to the place that He established for them."

It is extremely important to acknowledge this profound detail about Earth being formless because the volume of water in the beginning must remain consistent with our present-day volume of water. Understanding the Earth was "formless" is crucial for comprehending the validity

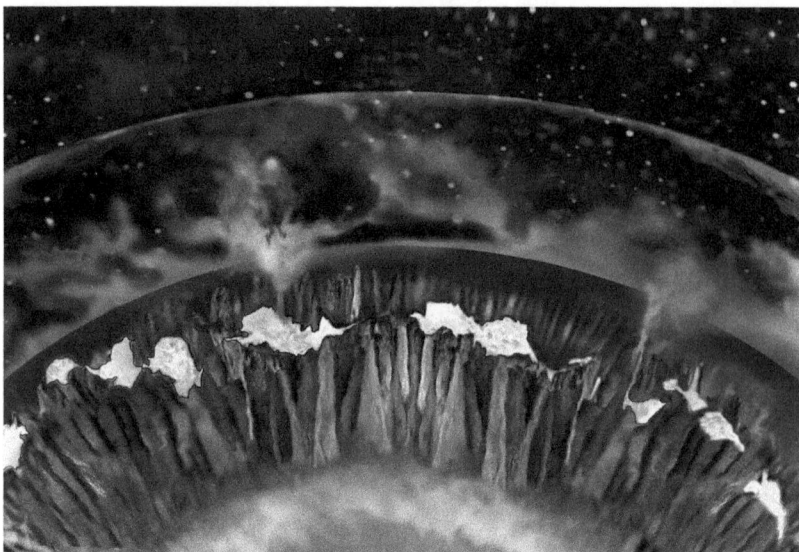

Steam Venting To Create The Vaulted Dome

and accuracy of Holy Scripture. Currently our planet is by no means formless. The highest mountain reaches upward of 29,000 feet. Scriptures tell us that in the beginning, the Earth was formless and its entire surface was covered with a vast body of fresh water. The volume of water on our planet has not changed. It is the same as it was in the very beginning. Earth's size has also remained constant, at about 197-million square miles. Taking into consideration that Earth was originally "formless," we may calculate that less than two miles of fresh water would have covered the entire smooth surface of Earth on day one of creation.

Water is the most common substance on Earth. It covers more than 70 percent of our planet. It also exists as a vapor in Earth's atmosphere. It is essential to life, greatly influences weather and climate, and plays a crucial role in many chemical reactions. Water has surprisingly powerful properties. It is a reactant and a catalyst, especially when heated, because hot water alters and increases the rate at which chemical reactions take place, and as a result it has power to facilitate an event of rapid change. At present only about 3 percent of Earth's water is freshwater. So, what process changed the fresh water into salt water? Where did all the water go when the dry land appeared? So many questions, and the answers were difficult to find, but God's Word provides a contextual model for us to persevere and understand the truth of His creation.

"All the water that will ever be, is right now."
Quote from *National Geographic*, 1993

For Truth is Truth

I am excited to share this story with you! A part of me wants to race forward and blurt out everything all at once, but it's probably best if we step back in time to 1993, the year when our journey of discovery began. The experience was so ominous that at times I was fearful. We knew we were up against established strongholds of deep deception. It was difficult to find anyone who would even consider the concepts we proposed. The impossibility of anyone understanding what we were saying was consistently brought to our attention, and we weren't too sure about how to say it anyway.

"For truth is truth though never so old, and
time cannot make false that which once was true."
—William Shakespeare

When we approached the pastor of our little Baptist church in Hornbrook, California, he chuckled and said, "Who are you to say anything? What is your educational background?" Fair enough. We told him it wasn't about our qualifications but about the truth of God's Word as it relates to the geology of Earth. His response was, "Well, I'm not qualified to comment on your work because without an understanding of geology I don't have a frame of reference to know whether or not what you are saying is true." I was disappointed and shocked by his response. All I could think was that even the pastor of our church did not have enough confidence in the absolute authority of God's Word to see it could be used as a frame of reference for understanding the observable features of Earth's surface.

He was not the only person of faith whose mind and heart were not open to the possibility of God's Word being precise, accurate, and true, even when applied to science and geology. I was disappointed. It was a quandary. People's willingness to accept God's Word as an exact, literal, representation of what we observe about His creation seemed to have been squelched. Indoctrination seems to have had a prevailing dominance over people's willingness to accept the Bible or to question science.

It wasn't always this way, for history reveals there were great minds during the scientific revolution. Brilliant

Bible-believing theists whose ideas and discoveries were influenced by God's Word. Universities like Oxford, Cambridge, and Yale began as monasteries, and people like Galileo, Sir Francis Bacon, and Isaac Newton became the founding fathers of the scientific method. They were not only inspired by the book of God's Word but their discoveries were also made by observing His works and studying creation.

After experiencing many trials, discouragements, distractions, and battles since 1993, I have concluded publication of this revelatory information has been fought and undermined on a spiritual level since the beginning. The fact you are reading this book now is a total miracle. But God! God has provided a way for each and every one of us to overcome by His power, love, authority, mercy, and grace. If not for my faith, and friends who understood my passion, this book would never make it to publication.

Our pastor's response was not discouraging; rather it provided fuel and increased our desire to validate and communicate more clearly what we perceived about the Word of God. With even greater fervor than when we first began, we diligently sought His Word for the truth about how our planet was formed.

I chose to spend some time at the University of Oregon Science Library. For several months I spent twelve hours a day, seven days per week with books piled high upon my desk. At times I had books stacked four feet tall as I diligently searched for evidence to substantiate the accuracy of God's Word. When I came across something that aligned with the biblical records of Holy Scripture, it was like finding gold! In the event what I was reading did not make sense or align with God's Word, it was discarded. Whenever I read something that did align with His Word, I cited the source. Using Holy Scripture as a siphon, straining through a great volume of scientific research, His Word became a tool for cultivating discernment.

God's Word provided an outline for viewing the geology of our planet from a completely new perspective. It established a framework and a solid foundation for comprehending volumes of complex information. As a result, logical, clear, and concise ideas began bubbling to the surface. One light bulb after another lit up as pieces of this grand puzzle began fitting together. The beginning of creation, what it was like to experience life here on Earth in ancient history, all of these seminal discoveries were in alignment with the exacting precision of His Word.

> "Heaven and Earth will pass away,
> but My words shall not pass away" (Matthew 24:35).

The Great Void

So where did all of the water covering the Earth's surface on day three of creation go when the dry land appeared? The original Hebrew word for "void" is *tohu wa bohu*. This same word is used in other biblical verses to describe emptiness like a vacuum. For example, Genesis 2:18 says, "It is not good for the man to be 'tohu wa bohu' (empty waiting to be filled), I

will make him a helper suitable for him." Our model of pre-
flood Earth is based upon the original Hebrew word for void.
Understanding the meaning of this word within the context of
it being a description of Earth's formation brings to light its
true significance. This very first verse of the Bible is describing an empty place beneath the Earth's formless surface. It
was an empty place waiting to be filled.

Gravitational forces generated by our planet spinning at
over 1,000 miles per hour would have caused water flowing at
Earth's formless surface to seek a path of least resistance. The
force of gravity would cause water flowing through cleft channels at Earths formless surface to swirl down like a spiraling

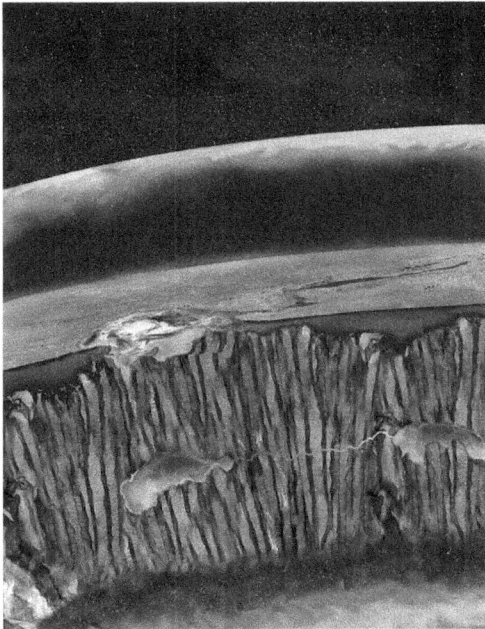

Gravity Forces Water Underground

cyclone at four strategic sinkholes, providing entrance to a vast cavity or recess of interconnected empty chambers beneath the surface into what is documented as the void. As a result of this dynamic process, on day three of creation the void became filled with flowing water and the dry land appeared.

Shallow freshwater seas would have remained within the parameters of sinkhole depressions. The same force of gravity pulling water underground would cause it to surge upward as a great artesian fountainhead in the Garden of Eden. From this fountainhead flowed four rivers forging their way through cleft rocky channels at Earth's smooth, formless surface. As water surged up, it would flow again to the strategic sinkholes where it once again drained into the underground chambers. This was a grand, hydrodynamic, gravity-fed, pump system encircling the globe, perpetuating and sustaining life just as is recorded in Scripture: "To the place where the rivers flow, there they go again" (Ecclesiastes 1:7).

Four Rivers

Geologic features at Earth's surface today reveal a remnant of cleft channels. They are a hint of where four rivers once flowed out of Eden. If you take the time to closely observe a map of the World's ocean floor, you will see a hint of where four rivers once flowed through cleft channels at Earth's formless surface before spiraling into the underground caves.

As I mentioned earlier, the caprock above this deep labyrinth of underground chambers was removed by an explosion. One that took place when all fountains of the great deep burst forth (Genesis 7:11). The cleft channel at the mid-ocean rift is a "Grand Canyon" now exposed at the ocean floor. It is the remains of an underground system of chambers. God inscribed this cleft channel upon the face of the deep when He laid the foundation of Earth as described by Job 38. And I believe it represents ground zero of a world-wide explosion taking place when all the fountains of the great deep burst forth (Genesis 7:11). This geologic feature reveals God's Word inscribed upon the face of the deep:

> "When He established the heavens, I was there;
> When He inscribed a circle on the face of the deep"
> (Proverbs 8:27, NAS).

The foundation of Earth at the ocean floor continues beneath the continents.

The four places on Earth today similar in composition to and meet up with the cleft channel at the ocean floor are:

1. The North American Cordillera
2. The African Rift Valley
3. The Verkhoyansk Mountain Range (A scientific article published by the West Virginia University (wvu) proposes the Verkhoyansk Mountain Range "is the mirror image of the North American Cordillera.")[1]

---

[1]"Structure and Thermochronology of the Verkhoyansk Fold-and-Thrust Belt: Implications for the Tectonic Evolution of the Sea

4. <u>The Ural Mountain Range</u> (a.k.a. Uralian Orogenic Belt) stretches 3,500 kilometers (about 2,175 miles) as an elongate mountain system extending from the Aral Sea to the Islands of Novaya Zemlya. The *Encyclopedia Britannica* exhorts that The Ural Mountain chain contains "slabs of the ocean floor [. . .] in the form of ophiolites"[1]

North of Iceland the cleft channel at the ocean floor pitchforks into three separate rifts within the arctic basin. This frozen region remains a mystery with many explorations in process but the continental cordilleras are an indication of their destinations. Mendeleev meets up with the North American Cordillera in Alaska, Gakkel Ridge leads to the Ural Mountain Cordillera in the Western region of Russia, and the Lomonosov Ridge leads to the Verkhoyansk Cordillera in Eastern Russia. These are three of the four places on Earth today revealing where four rivers flowed out of Eden. The fourth location is where the cleft channel at the ocean floor converges just south of the Red Sea and merges with the African Rift Valley; from there it diverges and departs on the right, then skirts beneath India before traversing the Pacific Ocean where, once again, the mid-ocean rift meets up with the North American Cordillera. "Now a river flowed out of Eden to water the garden; and from there it divided and became four rivers" (Genesis 2:10).

Okhotsk Margin of Eastern Russia," *pages.geo.wvu.edu/~jtoro/Research /Verkhoyansk.htm.*

[1] *Encyclopaedia Britannica*, s.v. "Uralian orogenic belt."

Fountain Vaulted Dome 4Rivers

After water covering the entire surface of Earth had filled the empty chambers and the dry land appeared, our planet truly was all land or "Pangaea." There were no great ocean basins or extremely high mountains, and a great fountainhead surged up from underground in the Garden of Eden, from where it divided and became four rivers.

The name of the first is Pishon; it flows around the whole land of Havilah, where there is gold. And the gold of that land is good; the bdellium and the onyx stone are there. And the name of the second river is Gihon; it flows around the whole land of Cush. And the name of the third river is Tigris; it flows east of Assyria. And the fourth river is the Euphrates. (Genesis 2:11-14)

With God's Word proving to be profoundly accurate about Earth's geology, I became increasingly curious about

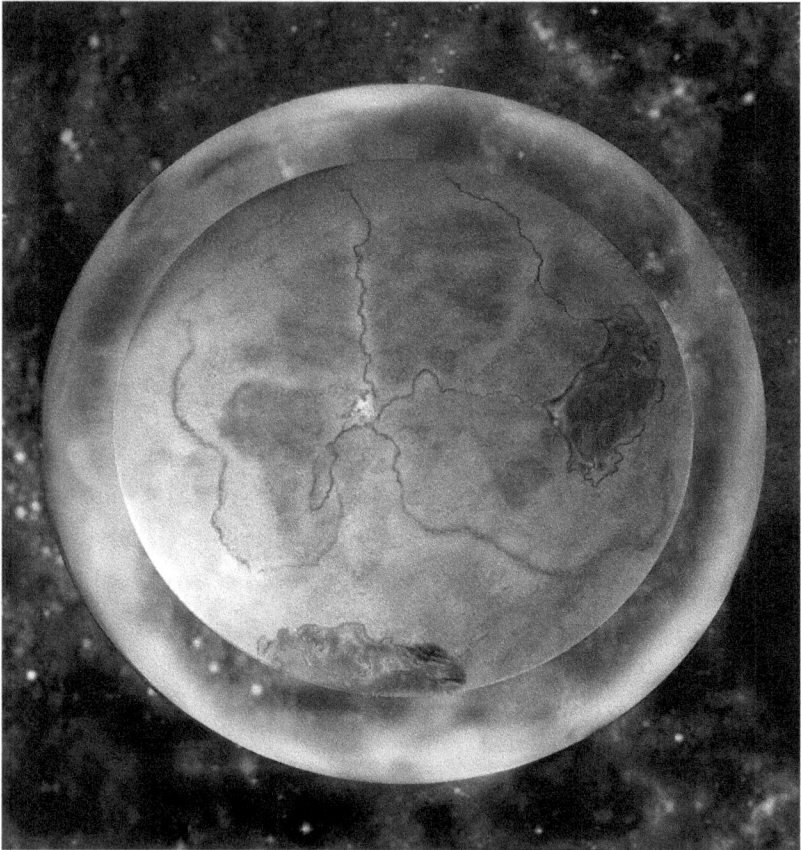

Four Rivers Shadow of Continents

the people who lived during these ancient times. I began to realize their stories were not legends or fairy tales, but actual accounts of their experiences, who they were, and how they lived. Our ancestors were not prehistoric but legendary people with a story to tell, just like you and I. I began to focus my attention upon unearthing evidence about their lives.

CHAPTER 3

THE ILLUSIVE GARDEN OF EDEN

"The most beautiful thing we can experience
is the mysterious.
It is the source of all true art and science."
—Albert Einstein

The Garden of Eden, from where four rivers flowed, was a Shangri-la of excellence. One can scarcely imagine how different the world was. Eden was an oasis of perfection, prepared to host the creation of the first man. Jesus was there in the beginning with God. "Then God said, 'Let Us make man in Our image, after Our likeness" (Genesis 1:26). When God saw that Adam was *tohu va bohu*, meaning he was empty and waiting to be filled, our Creator "caused the man to fall into a deep sleep, and while he slept, He took one of the man's ribs and closed up the area with flesh. And from the rib that the LORD God had taken from the man, He made a woman and brought her to him" (Genesis 2:21). 1 Corinthians 11:7 says man carries the brilliance of God's glory, and Psalm 104:2 describes our Creator God as being covered with light as if it were a garment of clothing. Therefore, Adam and Eve were radiant beings of light, created in God's image to live eternally and rule over the infinite resources of His creation (Genesis 1:26-28). Surprisingly, details about the countenance of Adam and Eve are preserved within the Chinese logogram for the word "glory." This is because ancient Chinese scribes originally constructed and developed written characters for the Chinese language based upon stories handed down from one generation to another about the mystical Garden of Eden. These stories became the building blocks of the Chinese language.

The Chinese character for the word "glory" is composed of symbols representing two persons "who look like fire with a bright and shining appearance"[1], and their light hovers above the tree of everlasting life.

Artistic representations for the tree of everlasting life have prevailed for millennia throughout many civilizations worldwide. Ancient artifacts discovered throughout the Sumerian, Harrapan, Egyptian, and Assyrian cultures show two winged, celestial guardians (angels) holding a type of fruit resembling the pinecone. The popular, modern, rendition of this symbol is the French "Fleur Des Les," which has a pinecone at the center. Today there are many artistic creations based upon

---

[1]C.H. Kang & Ethel R. Nelson, *The Discovery of Genesis: How The Truths of Genesis Were Found Hidden in the Chinese Language*, (Saint Louis: Concordia Publishing House, 1979), 52.

this pattern, but few recognize or understand its origin or significance as being a representation of God's promised gift of eternal life.

The ancient people of Egypt referred to the tree of life as the *axis mundi*. It became known as a cosmic pillar erected at the center of Earth marking the location where the first man, Atum/Adam, was created. Adam was known as "Atum." He was honored and revered as the father of humanity. The center of the Earth also became the sacred burial ground of Adam for it was at this sacred location where God breathed life into the dust of the Earth, creating the very first man in His own image and likeness. The genealogic records of Holy Scripture indicate Adam lived to be 930 years old. The average human lifespan in those days was about 847 years. Prior to the great flood of Noah, terrarium-like conditions were created by the vaulted dome, which not only shielded the Earth from harmful radiation but it also exerted an extra atmosphere of barometric pressure. Today it is known that additional atmospheric pressure reduces cellular oxidation, which is a natural byproduct of metabolism. This is why hyperbaric oxygen therapy (HBO) is used as a treatment for decompression sickness and other serious conditions in hospitals today. HBO therapy involves breathing pure oxygen in an environment where air pressure is two to three times greater than normal air pressure. Michael Jackson had a hyperbaric oxygen chamber. So how did modern science make this discovery? Perhaps this wisdom stems from the records of Holy Scripture. Amos 9:6 tells us "the one who builds His chambers in the heavens and has founded

His vaulted dome over the Earth [...] The LORD is His Name."

The idyllic conditions created by the vaulted dome allowed people and animals to live a very long time. Some of the exotic wild animals roaming the jungles grew to enormous sizes. For example, sea turtles, most fish, amphibians, lizards, snakes, and kangaroos have an indeterminate lifespan, which means they never stop growing for as long as they live. And these animals grew to become monstrous beasts under the vaulted dome. There was no terminal point in time they would stop growing throughout their entire life, and they lived a very long time. Beasts with tails the size of

Garden of Eden

old-growth cedar trees, creatures like the giant scaled reptile named "leviathan," were all subject to the authority of Adam and Eve (Genesis 1:26).

Legends of fire breathing dragons were passed down by ancestors of the Celtic, Greek, Chinese, and Roman cultures. One of the giant beasts was described by a man named Job:

> *Behold now Behemoth, which I made as well as you; He eats grass like an ox, Behold now the strength in his loins, And his power in the muscles of his belly. He bends his tail like a cedar; The sinews of his thighs are knit together. His bones are tubes of bronze; His limbs are like bars of iron (Job 40:15-18). In that day the Lord will punish Leviathan the fleeing serpent, With His fierce and great and mighty sword, Even Leviathan the twisted serpent; And He will kill the dragon who lives in the sea* (Isaiah 27:1).

The Great Deception

The Garden of Eden was the ultimate environment, but from the very beginning of creation, a great deception hovered over the Earth as *"darkness was over the surface of the deep; and the Spirit of God was moving over the surface of the waters"* (Genesis 1:2). This darkness over the surface of the deep may have been the spirit of Satan after he was cast down from heaven because he was already in the Garden of Eden when Eve stood next to the tree of knowledge of good and evil.

Lucifer was created by God as a light bearer. He had the seal of perfection, was full of wisdom and perfect in beauty (Ezekiel 28:12). He was the angel who had the honor of ushering prayers into God's most heavenly domain, into the holy of holies. But even before the first man was created, Lucifer desired to be worshiped and adored as if he were God. He believed he was more qualified to be God than the God who had made him.

> *How you have fallen from heaven, O Lucifer, son of the morning! [. . .] For you have said in your heart, I will ascend to heaven; I will raise my throne above the stars of God, And I will sit on the mount of assembly in the recesses of the north, I will ascend above the heights of the clouds; I will make myself like the Most High* (Isaiah 14:12-14).

The Bible tells us he was cast down from heaven: "I saw Satan fall like lightning from heaven." (Luke 10:18). He became the son of lawlessness, full of lies, trickery, and deception, the enemy of everything good, just, and pure. "Your heart was lifted up because of your beauty," lamented God, "You gave up your wisdom for the sake of your splendor" (Ezekiel 28:17). Lucifer's mission then, as it is now, was to be seated upon the throne of God and worshiped. According to the prophecy of Daniel 9:27 an ominous, fearful time is coming when Satan will deceive the world and seat himself upon the throne of the holiest place on Earth, in the temple of the Most High God (Mark 13:14).

As recorded by Ephesians 2:2, the father of lies and deception had dominion over the atmosphere. As one who has power to throw his voice over the air like a ventriloquist, the serpent spoke to Eve from within the tree of knowledge of good and evil (Genesis 3:4). The great deceiver enticed Eve to question what God had said about eating the fruit from this tree of knowledge: "You will not surely die. For God knows that in the day you eat of it your eyes will be opened, and you will be like God, knowing good and evil" (Genesis 3:4-5). As Eve began doubting what the Heavenly Father had told her and started believing the lies of Satan, her desire to be like God was all-consuming.

Adam and Eve had dominion over all of creation. They had the freedom to choose whether or not to obey God's edict, and they chose the one thing God had forbidden them to do. They ate from the tree of knowledge of good and evil. Not only were their eyes opened to knowledge but now they comprehended evil as well. Can you just imagine their predicament after disobeying God's commandment? Without the light that once covered them, they stood naked before God in the sacred garden. Their heavenly Father knew that if they ate from the tree of everlasting life in a sinful condition, they would forever remain in a state of corruption. As a result, they were evicted and blocked from re-entering the garden by two angelic beings with a flaming sword. This was God's way of protecting them. From this moment forward, all descendants of Adam and Eve would be vulnerable to deception and sin.

I can just imagine Adam and Eve yearning for restoration and longing for the close relationship they once had while walking and talking with their Heavenly Father in the garden. They worked and toiled, raised a multitude of children, and held on tight to the promise that one day God's plan of redemption would be fulfilled through the seed of Eve. Year after year they must have anticipated the birth of a child who would become their savior, God's promise of one who would have the power to forgive sin and restore them to righteousness. This child was not born during their lifetime, but Adam and Eve never stopped having faith in the covenant God had made with them. Retribution was ordained by God for the head of the great deceiver to be crushed by a descendant of Eve (Genesis 3:15). Not only would the serpent's head be crushed but a great price would be paid for sin and those who chose to believe in and accept God's gift of redemption, would be restored to righteousness, and once again have close fellowship with Him.

In spite of the humiliation, shame, and embarrassment experienced by Adam, he chose to repent of his sin. He wholeheartedly believed in God's promised gift of eternal life. Adam passionately spread the good news about God's covenant promise up until the moment he took his very last breath at 930 years old. Lucifer, on the other hand, remained on guard, waiting and watching to devour Eve's offspring as soon as the child was born: "The dragon stood in front of the woman who was about to give birth, so that it might devour her child the moment he was born" (Revelation 12:4).

The Father of Humanity

Adam was the very first man created in God's image and likeness. As the father of humanity, he was honored as a shepherd king—one to whom original wisdom was imparted by the Creator for guiding human souls back to eternity. Adam was a foreshadowing of Jesus Christ, Son of the Most High, Almighty God. Descendants of Adam who sought God's favor became wise old sages as they lived to be eight- or nine-hundred years old beneath the protective covering of the vaulted dome. These descendants were the patriarchal fathers who understood the physical and chemical nature of the mind, along with precise knowledge about the universe. Just like Adam, they had reverence for the fact they had been created in God's image and likeness. They did everything they could to uphold the sanctity of life.

Prior to the confusion and perversion of human history by sorcerers, pre-dynastic Egyptians remembered their ancestors as being the "holy ones." Stone tablets carved with their names describe these ancestors as being "the righteous sons of God." Column one of the Turin King List reveals seven kings and speaks of pre-dynastic Egypt as being the domain of the *ntr*, a term meaning quite literally, the "gods." The Palermo Stone, the Abydos Tablet, and the Turin King List begin with an enumeration of gods that ruled Egypt 5,000 years ago. Now according to the genealogical records of Holy Scripture, this was before the great flood of Noah, which took place about 4,386 years ago. Most historians believe these were mythical gods, but biblical timelines and carvings within stone tablets indicate some of these people were the noble shepherd kings whose lives were extremely long. They had an awe-inspiring radiance as they ruled with the light and presence of the Lord and carried forward sacred wisdom imparted to them by God Almighty. They were the patriarchal fathers who maintained knowledge of the promised gift of eternal life. Their names read in succession spell out God's plan of redemption.

Names of the Pre-Flood Patriarchs:
Meaning & Significance

To elaborate upon the significance of names given to Pre-Flood patriarchs, in sequence their names spell out God's promise of redemption beginning with Adam, the first created man. Then Seth, Enosh, and Kenan would be born with names meaning Appointed, Mortal, and Sorrow; foretelling of Jesus being Appointed to Mortal Sorrow. He was Blessed God manifest in the flesh who Came Down to Earth, Teaching, before His death on the cross, resulting in The Despairing, leaving us with the Comforter.

The ancient book named *Cave of Treasures* records that all of Adam's offspring gathered together and were blessed by him before he died. Seth was 800 years old when his father passed away at the ripe old age of 930 years. I invite you to imagine how experienced and intelligent you and I could be with an opportunity to live this long? According to his father's instructions, Seth prepared Adam's body; "embalm

Name	Meaning
Adam	Man
Seth	Appointed
Enos/Enosh	Mortal
Cainan /Kenan	Sorrow
Mahalalel	Blessed God
Jared	Shall Come Down
Enoch	Teaching
Methuselah	His Death Shall Bring
Lamech	The Despairing
Noah	Comfort

Names and Significant Meaning

me with myrrh, and cassia, and stakte, and deposit my body in the Cave of Treasures."[1] Embalming his father's body after an unexpected death likely had a profound influence

---

[1]"The Cave Of Treasures," https://sacred-texts.com/chr/bct/bct04 .htm, 73.

upon Seth because he most likely believed God's promise of redemption would have been made manifest during his lifetime. The death of his father Adam must have had an immense impact upon all who knew him. It was a great sacrifice, and as a result, Adam's death became a foreshadowing of the fulfillment of God's promise. I mean it could've gone one of two ways. Seth could have given up and would no longer believe in God's promise or he would choose to stay the course established by his father. The experience of his father's death apparently made Seth's belief in God's promise of eternal life even more certain because he continued to share his father's knowledge of mummification. The performance of this sacred burial was to "preserve" the message of God's promised gift of eternal life. It was all about perpetuating hope for humanity being restored to a "sinless" condition and inheriting a transformed body to experience the gift of eternal life.

Memorialization of God's Promise

The memorialization of Adam's life purpose was hallowed for generations. He was the first man to embrace and communicate God's message of salvation and was known as *Asar-Sa*, meaning the "Shepherd King":

> *Among the ruins of ancient Egypt are the records of a bearded, long-haired Savior, whose resurrection from the dead influenced generations of followers to believe in life after death.*

He was called "Shepherd," "King of Kings," and "Lord of Lords," and he was depicted in art for thousands of years with the ankh cross of eternal life. The first King of Egypt, he was recorded in myth as the recurring "Phoenix" bird who appears perennially in the East to herald each new age. As Judge of the Dead on the Day of Judgment—the position Christ later held—he enforced the sacred law. Every Pharaoh dressed in his image (a bearded shepherd with long hair—identical to Christ) while sitting on his throne [...] The deceased were entombed under his likeness to gain immortality through him.[1]

Like Seth, Egyptian embalmers prepared the deceased bodies of Kings so they too would be able to inherit the gift of eternal life. This sacred ritual took seventy days to complete. But Adam's teachings about mummification became monetized, and as a result, it was no longer a sacred practice but a lucrative business. Resurrection and the promised gift of eternal life were embraced by those who sought truth, but there was a powerful deception at work before the flood of Noah. Many concepts and symbols became twisted and corrupted for the purpose of serving false gods, even though, in the beginning, knowledge of the one true God who created Adam was common. If the biblical account about man being created from the dust was true, wouldn't you expect that such a man as Adam would have been revered by those

---

[1] E.A. Wallis Budge, *Osiris and the Egyptian Resurrection*, (Martino Fine Books 2019).

who knew him? Nothing would be more natural than for his descendants to have created a permanent memorial in honor of his sacred wisdom. American theologian Joseph A. Seiss wrote in 1877 that "the Jews up to the Savior's time had a cherished tradition that the Great Pyramid of Giza was built before the flood."[1]

Adam was also honored with the name "Atum-Ra," a symbol of the sun representing our righteous Creator in heaven who rises each day with healing in His wings. "But for you who fear My name, the sun of righteousness will rise with healing in its wings, and you will go out and leap like calves from the stall" (Malachi 4:2). Adam intimately knew his Father from the moment he was created. After repenting of sin, he was in obedience to the spirit of righteousness and feared God. So yes, the spirit of God known as "Ra" by the pre-flood people living in ancient Egypt, Atum was their connection to know and understand the promise of the Creator God through the Holy Ghost. This is the legacy of Adam who was created in the image and likeness of God.

---

[1] Joseph A. Seiss, *A Miracle in Stone: The Great Pyramid*, (1877), 172.

CHAPTER 4

CREATED IN GOD'S IMAGE

It is believed by some that the statue of the Sphinx adjacent to the Great Pyramid of Giza was originally built as a representation of the living image of God. It was built in honor of Adam/Atum being the first human being created in God's image and likeness (Genesis 1:27). The word "sphinx" is Greek, but the Egyptian name for this statue was "Shesep-Ankh," meaning "living image." This ancient artifact is a symbol of supremacy, royal strength, and majesty. It was created to house the spirit of God, the One who watches over and protects the dead. The One whom the Egyptians refer to as Ra.

It is not hard to imagine the profound impact Adam's death would have had upon those who honored him as the first man created in God's image and likeness. His passing was an unforeseen event in light of God's promise to crush Satan's head with one of Eve's offspring. Even though Adam died before God's promise was fulfilled, he completely trusted in God's gift of redemption, which is a testament to the unfailing love Adam had for his Father/Creator, God Almighty. Monuments founded in memory of Adam's passing indicate what a huge impact he had upon the world during his lifetime. The head of the Great Sphinx upon the plateau of Giza just so happens to be one and the same with the funeral mask of Adam/Atum, the Shepherd King.

Early Jewish writings say the mummified body of Adam was brought to Annu and placed in the sarcophagus within the Great Pyramid, at the middle of the Earth. The Egyptian Book of the Dead *says the mummified body of Atum resided in the House of Annu within the castle of the pyramid, the Great Pyramid. It's the same story.*[1]

Research by Wallis Budge, Schwaller de Lubicz, Allen Austin, and many others collectively asserts, "the Great Pyramid is the original Mount Zion." The memorial place of Adam, the first man the Father of Humanity. It was also known as the ancient city of Annu located in the middle of the Earth. Author Allen Austin says, "I am convinced it was built in the antediluvian world, long before the Egyptians existed. It was built by the children of Seth, a line of shepherd kings of the master order."[2] Doctor of Theology Ken Johnson reveals the misunderstood timeline of this "pre-flood monument" with this explanation:

Other examples of secular historians confusing history are: Ceops (meaning "lord of light") created the great pyramid. The name exists in pre-flood king lists and the fourth dynasty

---

[1] Allen Austin, *The Middle of the Earth*, (Xulon Press, 2011), 76.
[2] Allen Austin, *The Middle of the Earth*, (Xulon Press, 2011), 136.

*Egyptian king list. This pre-flood monument's history is cred-
ited to the wrong Ceops.*[1]

One way this deception came about, is recorded by Daniel
7:25, who tells us the great deceiver would have the audacity
to think he could change the record of time as documented
by Holy Scripture. I will venture to say he's done a very good
job of it. The following chart reveals the timeline of human
history according to genealogical records of the Bible.

One piece of evidence for the corruption of our history's timeline was recently brought to light with the discovery of erosion
and weathering by water at the base of the sphinx adjacent to
the Great Pyramid of Giza. There are channels of erosion two
feet deep within its limestone sides, indicating it must have been

PRE-FLOOD BIBLICAL TIME LINE		
EVENT	BIBLE VERSE	YEAR
ADAM CREATED	GEN 5:1	1
SETH BORN	GEN 5:3	130
ENOS BORN	GEN 5:6	235
CAINAN BORN	GEN 5:9	325
MAHALALEL BORN	GEN 5:12	395
JARED BORN	GEN 5:15	460
ENOCH BORN	GEN 5:18	622
METHUSELAH BORN	GEN 5:21	687
LAMECH BORN	GEN 5:25	874
NOAH BORN	GEN 5:28, 29	1056
FLOOD OCCURRED	GEN 7:11	1656

---

[1] Ken Johnson, *Ancient Post-Flood History*, (2010), 108.

built prior to "the arid desert conditions found in the region during the last four thousand to five thousand years. Rather, the observed weathering resulted from rain, precipitation, and water runoff, and sufficient precipitation was available only during the pre-Sahara conditions, prior to circa 3000 BCE."[1] This new evidence for the age of the sphinx mysteriously places the time of its construction to the time of Adam's death, about 3,074 BC, according to the chronology of Holy Scripture.

John Anthony West postulates the anomalies and contradictions of recent archeological and geological surveys only substantiate the vast antiquity of the Sphinx: "Of course this means totally rewriting the accepted chronology of the evolution of human civilization."[2] Scientists from a Stanford University Research team agree other Egyptian monuments built with stone similar to the Sphinx have not suffered the same type of erosion. Their research led author John Anthony West to say, "Meanwhile, the erosional channels cut into the Sphinx are nearly two feet deep," and the evidence is "irrefutable: the Sphinx and its temple complex are vastly older than all the rest of Dynastic Egypt, the weathering is due to the action of water—of a flood."[3]

---

[1] Robert Schoch, *Forgotten Civilization: The Role of Solar Outbursts in our Planet*, (Rochester: Inner Traditions, 2012), 223.

[2] John Anthony West, *Serpent in the Sky*, (Wheaton: Quest Books, 1993), 223.

See also: Robert Schoch, "The Great Sphinx of Egypt (at Giza)," https://robertschoch.com/sphinx.html.

[3] John Anthony West, *Serpent in the Sky*, (Wheaton: Quest Books, 1993), 207-208.

Prior to the great flood of Noah, wouldn't our very wise ancestors have attempted to preserve their story for all to see and know who the Creator was, who He is, and is to be? Especially for those who live during the last days as they witness mockers of truth becoming overwhelmed with terror as a great refuge of lies are swept away: "And I will make justice the measuring line, And righteousness the level; Then hail shall sweep away the refuge of lies, And the waters shall overflow the secret place" (Isaiah 28:17).

A pervasive deceit has prevailed over the ages distorting the true significance of monuments like the Great Pyramid of Giza, along with many other ancient records etched in stone. But these artifacts have withstood the tides of time. The original Latin word for pillar was "pyramis," a word meaning "pillar of fire and knowledge." Another Latin word is pyra-mid, meaning "fire within." In the Bible it is prophesied that in the last days, the Lord will make himself known through a monumental pillar in Egypt; according to Isaiah 19:19, "In that day there will be an altar to the LORD in the center of the land of Egypt." The Great Pyramid of Giza is located at the center of the Earth and was built of stone to withstand a great flood. It was built with megalithic blocks that look as though they were carved with laser-like precision. This giant structure aligns so precisely with the cardinal seasons that eight sides, and not four, can be seen as indentations. They are revealed by light from the sun as it shines from very specific angles during the spring and fall equinoxes. It is rather obvious, builders of the Great Pyramid of Giza calculated and strategized its dimensions using spherical trigonometry, because they must have considered the circumference of Earth for any of their measurements to be precisely in alignment with the rising sun at the beginning of each one of the four seasons. Their ancient measurements are more accurate than our current Greenwich Mean Time (GMT).

More marvelous facts:

1. The Great Pyramid of Giza approximates the squaring of the circle with better accuracy than the Kepler triangle.
2. A circle drawn around the perimeter at the base of the Great Pyramid has a radius equal to its height with an error point as low as 0.04 percent . . . "Or, that the vertical height of that Pyramid was to the length of one side of its base, when multiplied by 2, as the diameter to the circumference of a circle; i.e. as 1 : 3.14159."[1] This is a well-known mathematical formula referenced as the golden ratio, pi (℗), the Fibonacci sequence, or the divine proportion. Measuring the circumference of a circle and dividing its numeric value by the diameter consistently results in 3.1419, no matter what size the circle.
3. The numeric value for the speed of light within a vacuum is 29.9792458 m/sec. Using this number as a northern coordinate results in finding the latitudinal location of the Great Pyramid of Giza on a map.

Smooth, white, casing stones once covered the entire structure of the great pyramid. According to several Arab and

---

[1] Charles Piazzi Smyth, *Our Inheritance in The Great Pyramid*, (London: W. Isbister & Co., 1874), 15.

Greek historians the mysteries of astronomy, science, geometry, and medicine were etched into the pearly white casing stones covering this monument. With laser-like clarity and intensity our history was recorded, inscribed, and preserved. Herodotus personally witnessed the etched writings during his Egyptian visit around 450 BC and described the pyramid as being covered with "polished stone, with figures carved on it." He also was amazed at how precisely the polished stones were "fitted together with the utmost care."[1] Seiss believed the outer casing stones of the great pyramid were encoded with advanced knowledge of the physical universe and the course of human history.[2] After being loosened by a fourteenth century Earthquake, the outer casing stones of the Great Pyramid of Giza were pillaged to build fortresses and mosques in the nearby city of Cairo, resulting in the history of our ancestors being left in the deteriorated step-stone condition we see today. Prior to the defacement of this monumental Seventh Wonder of The World, if you were to look down from outer space upon its polished limestone walls reflecting the sunlight, it would have looked like a brilliant star shining upon the Earth.

One of the most fascinating highlights upon my journey of discovery was the delivery of a very old book. It was written in the 1800s by Piazzi Smyth, who was ordained as the "Astronomer Royal" for Scotland during the years from 1846 to 1888. As it turned out, this book began with John Taylor

---

[1] "Herodotus Book II: chapters 99-182," https://penelope.uchicago.edu/Thayer/E/Roman/Texts/Herodotus/2b*.html.

[2] "Masoudi Manuscript Document 9575," British Museum.

who just prior to his death commissioned the completion of his work to Piazzi Smyth. He carried forward the torch of Taylor's investigations and research on the Great Pyramid. After Piazzi Smyth accepted this responsibility, and years of meticulous measurements, he published the book "Our Inheritance in the Great Pyramid" and dedicated it to his friend John Taylor.

Smyth employed a variety of surveying and astronomical instruments to obtain measurements of the mighty monument, ones far more exacting than had ever been attempted before. He documented precise and strictly provable data to demonstrate the radical difference between the Great Pyramid and every other pyramid in Egypt.

> *In fact, these measures tend to establish that the Great Pyramid, though in Egypt, is not, and never was of Egypt—that is, of belonging to, or instructing about Pharaonic, idolatrous, and chiefly Theban Egypt. Also, that though built in the earliest ages, far before written history, the Great Pyramid was yet prophetically intended—by inspiration afforded to the architect from the one and only living God, who rules in heaven, and announced vengeance against the sculptured idols of Egypt (Ezekiel xxx. 13)—to remain quiescent during those earlier ages; and only, in a manner to come forth at this time to subserve a high purpose for these latter days. That it, the Great Pyramid, was never even remotely understood, either by the Egyptians, or any other branch of the Cainite and anti-Israelite family of nations.[1]*

---

[1] Charles Piazzi Smyth, *Our Inheritance in The Great Pyramid*, (London: W. Isbister & Co., 1874), ix-x.

This work, begun by John Taylor and continued by Piazzi Smyth, was published in 1880 as a 600-page book with amazing illustrations and calculations. In the preface to his book Smyth describes the prophetic significance of the Great Pyramid as a preservation of human history within a monument of stone containing **extremely valuable information for a very unique and future generation** . . . "a message conveyed not by the use of any written language, whether hieroglyphic or vulgar, but by aid of the mathematical and physical science of modern times"[1] (p ix). A message that will be understood by all nations "when the fullness of prophetic time, in a science age, has at last arrived" (p x). Smyth sums up the preface of his book by saying the collective work of their discoveries, along with works yet to be cited, will absolutely prove this monument of old was built by supernaturally inspired men, prophets of the living God. "The Egyptians did not exist in the antediluvian world. They entered the world scene after the great deluge, thousands of years after the creation of Adam (Atum)."[2]

To this day, modern man has been unable to build anything like the Great Pyramid. In order to account for this awesome display of wisdom on the part of our ancestors, how can we fail to acknowledge God's hand in this creation? As indicated by Isaiah 19:19 the intricate details and precise alignment of this grand construct may have been intended for a last days

---

[1] Charles Piazzi Smyth, *Our Inheritance in The Great Pyramid*, (London: W. Isbister & Co., 1874), ix.

[2] *The Middle of the Earth*, Allen Austin, p 38

revelation when the Lord will make Himself known. "In that day there will be an altar to the LORD in the midst of the land of Egypt."

A Corruption of Knowledge

What if the Great Pyramid of Giza truly was constructed prior to the flood of Noah? What if the giant Sphinx adjacent to this construct does portray the "living image" of God who created Adam in His likeness as the father of humanity, ? Could it possibly be that our ancestors living prior to the Earth-shattering event of Noah's flood were attempting to send future generations a message to save souls? It seems to me they were passionate enough to record their story in such a way that it would remain far into the future. Perhaps their message was intended for a future generation who would be challenged to survive a time that paralleled their own? A time when evil, sorcery, and deception would once again overtake the world as it did during the days of Noah.

I believe the significance of this great monument may be that it has preserved the truth about the roots of human history, and it is a testament to the significance of our Creator, who is the Alpha and the Omega. The One who from the beginning of time, prepared a way for the redemption of lost souls. Our Father who art in heaven paid an extremely high price to cover our iniquity with the sacrifice of His only begotten Son, and those who choose to believe in "The Truth" will be restored to righteousness. They will be brought into fellowship with the

Light of the World, who is our Lord and Savior Christ Jesus. The last generation who believes and worships Him in spirit and in truth are not only heirs of a promised redemption; they will not experience death. Christ returns for His bride who is caught up in the twinkling of an eye. Christ's bride is made up of believers who wait and watch for His return. "Then we who are alive and remain shall be caught up . . ." (1 Thessalonians 4:17).

A dire warning about the twisting of knowledge leading to a final and great deception is emphasized by author Gary Wayne, who has meticulously followed and documented a history of tainted bloodlines from the beginning of creation until now.

I cannot underscore this point aggressively enough; the heavenly sciences and knowledge were corrupted, engineered, and preserved, all to delude this unsuspecting generation and get it to rebel away from God and into destruction. This knowledge has been cleverly hidden and carefully guarded within a secret trust until its time of destiny. References to this knowledge have remained unaccountably alive and well in myths and legends from all cultures around the world, all waiting for the Antichrist and his reign of terror that is to come. Templars believe that indeed this Golden Age once existed and will once more, in the New Age of light, Aquarius, and the end times.[1]

The great deceiver knows his time is short and feverishly works to captivate the minds of men. "See to it that no one deceives you. For many will come in My name claiming, 'I

---

[1]Gary Wayne, *The Genesis Six Conspiracy*, (Trusted Books, 2014), 681.

am the Christ' and will deceive many" (Matthew 24:5). Jesus is "The Truth" who will guide our every step if we trust in Him and know who is "the lie" and the great deceiver.

I encourage you to absolutely have no doubt but complete confidence in "the mysterious and hidden wisdom of God, which He destined for our glory before time began" (1 Cor 2:7). Do not be deceived for "we have not received the spirit of the world, but the Spirit who is from God, that we may understand what God has freely given us. And this is what we speak, not in words taught us by human wisdom, but in words taught by the Spirit, expressing spiritual truths in spiritual words. The natural man does not accept the things that come from the Spirit of God. For they are foolishness to him, and he cannot understand them, because they are spiritually discerned" (1 Corinthians 2: 12-14).

Perhaps the true significance of these monumental achievements by our pre-flood ancestors was meant to be withheld for an appointed time, released for the specific purpose of strengthening hearts about to fail. No one will know the day or the hour for the great unveiling of Christ, what is known as the apocalypse. Perhaps during the great tribulation eyes will finally be opened and the pre-flood patriarchal fathers will no longer be thought of as myth but as a foreshadowing of the Son of God. Perhaps when people realize they are in the middle of a great deception they will be motivated to believe God is who He says He is. That He has done and will do what He says He will do. Psalm 19 tells us:

The heavens declare the glory of God; the skies proclaim the work of His hands. Day after day they pour forth speech; night after night they reveal knowledge. They have no speech, they use no words; no sound is heard from them. Yet their voice goes out into all the Earth, their words to the ends of the world.

The Bible says God created the stars and calls them all by name (Isaiah 40:26). Today there is a great deception at work just as there was prior to the flood of Noah.

Because of corruption people began worshiping the stars and planets themselves, not the One who made them. Today we practice a corrupt form of astronomy and astrology. The ancients understood the significance of the stars, and the star of Bethlehem is confirmation. Even at this time the deviation was animate. It was the wise Magi who retained the pure view and recognized the paramount event in human history, the birth of Christ. What better lesson is there?[1]

Genesis chapter three ends with Adam and Eve being banned from the Garden of Eden. And only a few chapters later in Genesis six an ominous seed of corruption is brought to light. Author of "The Genesis 6 Conspiracy" poignantly warns, "The hidden hand of hubris haunting humankind's history must be revealed, along with its duplicity and its surreptitious serpentine organization. Its sins against humanity must ultimately be reconciled with our world and creation— all through free choice."[2]

---

[1]Allen Austin, *The Middle of the Earth*, (Xulon Press, 2011), 211.

[2]Gary Wayne, *The Genesis Six Conspiracy*, (Trusted Books, 2014), 688.

CHAPTER 5

CAIN KILLS HIS BROTHER

To get a clear picture about who these biblical characters were, I set out to better understand their lineage
and explore archeological discoveries of historic locations where they may have dwelt. Genesis chapter five
"is the book of the generations of Adam. In the day when
God created man, He made him in the likeness of God"
(Genesis 5:1). At first glance Genesis chapter five may
seem boring, but upon closer investigation I realized how
extremely important this chapter is because it provides an
accurate accounting of the number of years from the time
of Adam's death until the great flood of Noah, 726 years.
Not only is Genesis five extremely valuable for measuring
the genealogical record of time, but the names of Cain and
Seth's lineages are far too similar for them to be coincidental. The parallel names outlined in this chapter are a
strong indication of the stage being set for a spirit of confusion to prevail.

Seth's Lineage	Cain's Lineage
Adam	Adam
Seth	Cain
Enosh	Enoch
Kenan	Irad
Mahalalel	Mehusjael
Jared	Methusael
Enoch	Lamech
Methuselah	Jabel, Jubal, Tubal, Naamah
Lamech	
Noah	

Lineages of Seth and Cain

I believe one of the greatest benefits of studying Genesis chapters five and six is to cultivate discernment and establish awareness of a spurious bloodline that has woven its way throughout human history up until our present day. It was during this rather short period just prior to the flood that there were saints who followed Seth's lineage and deviants who followed in the footsteps of Cain. The following chart reveals the lineal descent and paralleling names of the two brothers and how a mistake in identity could have easily led to deception. Therefore, tracing these lines of ancestry is extremely important.

As you may have already heard, Eve's first son, Cain, cultivated the Earth as a farmer. His younger brother, Abel, was gifted with the knowledge of raising and herding sheep. These two brothers were brought up to honor God with the gifts of their labor. But one day Cain became enraged with jealousy because God chose to honor his brother's lamb as an acceptable sacrifice. Cain took offense, thinking his harvest from the garden was of no comparison. A spirit of envy began to dominate his consciousness as Cain dwelled upon the blessing his brother had received. This was a privilege he coveted and felt entitled to.

Cain's resentment grew despite God's attempts to counsel him about mastering his anger. He was unable to forgive his brother. As his hatred grew, a doorway opened for the powers of evil and deception to have dominion over Cain's heart, mind, and soul (Genesis 4:6). Cain's envy and covetousness, his inability to share in the joy of his brother's blessing, became an open invitation for the deceiver to trap him in a downward, spiraling vortex of disgust and hatred. Cain was so enraged that a different personality consumed him with a spirit of hatred, resulting in the murder of his very own brother.

And the LORD said to Cain, "Where is Abel, your brother?" And Cain replied, "I do not know. Am I my brother's keeper?" and God said, "What have you done? The voice of your brother's blood is crying to Me from the ground" (Gen 4:10). God knew what Cain had done and where Abel was. His question was intended to prompt Cain to self-reflect and admit his wrongdoing. This is the first step toward being forgiven, receiving guidance, and being restored to righteousness. But Cain responded with a lie. As a result, he was cast away from God's presence, and Cain feared being put to death for what he had done. He pleaded with God for protection, and God was merciful. He marked Cain with a sign of vengeance, as a warning for anyone who might attempt to punish or kill him. God was just. He let Cain know he would suffer the consequence of his actions. "You are under a curse and driven from the ground, which opened its mouth to receive your brother's blood from your hand. When you work the ground, it will no longer yield its crops for you. You will be a restless wanderer on the Earth" (Gen 4:11-12). Cain believed he was condemned to wander the Earth forever as a vagrant, but in due time, he settled in a place called Nod, a word meaning "to wander."

The first century Jewish historian Flavius Josephus states that "Cain did not accept his punishment in order to 'make amends,' but to increase his wickedness." Josephus goes on to say Cain prospered by overpowering and stealing from his neighbors. As a result, he acquired great wealth and became a successful trader of goods. Cain and his descendants took comfort in their worldly possessions. It is recorded that his great, great, great, grandsons, Tubal and Jubal Cain, were gifted with the talent of fabricating musical instruments, such as the harp and organ, along with **weapons** made from bronze and iron (Gen 4:21-22). The creation of these metal weapons and instruments would have required expertise with smelting, forging, and shaping, along with knowledge about where to find bronze and iron to extract it from the ground.

Cain's ancestors took great pride in their unique abilities, and eventually their skills were used to capitalize on the creation of trinkets and idols. The *Book of Jasher*, or *Sepher HaYashar* in Hebrew, is referred to in Joshua 10:13 and 2 Samuel 1:18 as a source of additional historical information to the Old Testament. Literally translated, the title of this book means "Book of the Upright." Jasher paints a colorful description of this point in time.

In those days the sons of men made images of brass and iron, wood, and stone, and they bowed down and served them. And every man made his god and they bowed down to them, and the sons of men forsook the Lord all the days of Enosh and his children.[1]

Cain's Ancient City of Erech

"Then Cain went out from the presence of the Lord and settled in the land of Nod, east of Eden . . . and he built a

---

[1]Jasher 2:4-5

city, and called the name of the city Enoch, after the name of his son" (Genesis 4:16-17). The city built by Cain had great walls for protection because he feared retribution after killing Abel. I began searching history books along with archeological evidence to confirm the story of Cain. According to recent archeological discoveries the ancient City of Enoch built by Cain is most likely buried beneath the ruins of a post-flood establishment known as Uruk—modern-day Warka, Iraq. Uruk was also known in the Aramaic language as Erech, which is mentioned in Genesis 10:10. It is interesting to note that modern scholars still regard Uruk as the world's first city. And sure enough, this construct reveals a great thickness of walls that may have been built to protect Cain. Arthur Custance explains that cuneiform tablets from Assyria refer to a City named Erech (Uruk), which when pronounced with the inflection of "N" for the letter "R" translates to the name Enoch (Erech).[1]

It was customary to build on top of the foundations of earlier sacred sites. Inscribed clay tablets and cylinder-seals unearthed from ancient cities are intrinsic to understanding what life was like during this early Genesis period. These relics record the names of people and places not documented in biblical narratives until thousands of years later. British archaeologist Leonard Woolley discovered through his excavations at Ur that there were generations of constructions built one on top of the other with prior generations. This

---

[1]Arthur C. Custance, "The Original Speech of Mankind," https://custance.org/Library/Volume6/Part_V/Chapter2.html, 8.

was done in order to honor the sacred dwelling places of their revered ancestors. The City of Erech had an excavation depth the height of a five-story apartment building. "No less than 18 separable strata, of which levels 5-18 were prehistoric; the total depth of the excavation was about 19-meters, of which the bottom meter was virtually sterile."[1] Perhaps layers covering the pre-flood city of Erech are an indication of life being blotted out during the great flood of Noah (Genesis 6:7). Something to think about!

The Sumerian King List is a stone tablet estimated to be 4,200 years old. It has inscriptions recording that the First Dynasty of Uruk (Erech) was centered within the temple precinct E-Anna, where "twelve kings are said to have reigned for 2,310 years. Among these kings was Gilgamesh."[2] The Epic of Gilgamesh is an ancient Mesopotamian record of the flood. It was most likely recorded by the surviving ancestors of Noah and is one of many renditions containing descriptive details about this catastrophic event. The information age is bringing legends of biblical history to life as the geology of Earth, ancient artifacts, and historic records give us a glimpse into the history of humanity. Genesis chapter four chronicles the story of Cain killing his brother, Abel, and the start of Cain's lineage. It ends with Seth being born and Eve saying, "God has appointed me another offspring in place of Abel; for Cain killed him."

---

[1] *Cambridge Ancient History*, 361.

[2] Jack Finegan, *Light from the Ancient Past*, (Princeton Univ. Press, 1946), 32.

Chapter four also records the birth of Seth's first son, whom he called Enosh.

Eve's Third Son, Seth

Seth was Eve's third son, whom God appointed in place of her son Abel, who was murdered by Cain. Eve's third son, Seth, produced a lineage of family members with very different qualities from Cain's offspring. His descendants were historically referred to by the Jews, Christians, and Islam as being "the righteous ones," a virtuous people who called upon the name of the Lord.

The righteous ones understood God as a being of light who made Himself known to them through His presence and by His creation. Antiquities of the Jews by Flavius Josephus says that "Seth's descendants were inventors of that peculiar sort of wisdom which is concerned with the heavenly bodies, and their order [. . .] they inscribed their discoveries into great pillars of stone so their inventions would not be lost without becoming sufficiently known."[1] Flavius Josephus goes on to say that Adam predicted the world was to be destroyed at one time by the force of fire and at another time by the violence and quantity of water, so they made two pillars, the one of brick, the other of stone: they inscribed their discoveries on them both, that in case the pillar of brick should

---

[1] Josephus Flavius, *Antiquities of the Jews*, (Blacksburg: Unabridged Books, 2011), 67.

be destroyed by the flood, the pillar of stone might remain, and exhibit those discoveries to mankind, as well as inform them that there was another pillar of brick erected by them. Now this remains in the land of Siriad to this day. "In that day there will be an altar to the LORD in the center of the land of Egypt, and a pillar to the LORD near her border" (Isaiah 19:19).

The Great Pyramid of Giza is a pillar of stone located within the very center of the land of Egypt. It lies in the middle of Egypt and in the middle of lower and upper Egypt. The north-south axis (31 degrees east of Greenwich) is the longest land meridian, and the east-west axis (30 degrees north) is the longest land parallel on the globe (Our Inheritance in The Great Pyramid, Piazzi Smyth).[1] And sure enough, as recorded by Isaiah 19:19, Gobekli Tepe, with its walls made of brick, is located along what used to be the northern border of Egypt as shown on this fourteenth century BC map by Andrei Nacu, Jeff Dahl—Own work based on Egypt 1450 BC.[2]

Gobekli Tepe was discovered in the 1990s by Klaus Schmidt. In English, the name of this place means Potbelly Hill and represents an umbilical cord connecting man to the inception of human history.[3] Archeologists estimate that this site is thousands of years older than the Great

---

[1] Allen Austin, *The Middle of the Earth*, (Xulon Press, 2011), 125.

[2] Egypt NK edits.svg, https://commons.wikimedia.org/w/index.php ?curid=4335117.

[3] Ronnie Jones III, s.v. "Göbekli Tepe," *World History Encyclopedia*, (May 2015), https://www.worldhistory.org/G%C3%B6bekli_Tepe/.

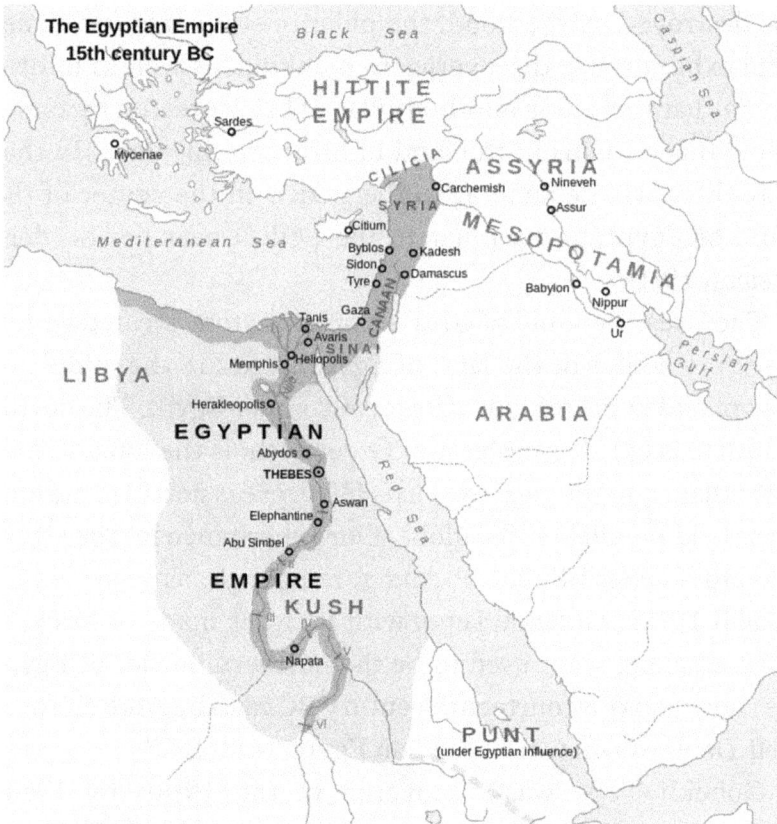

Map of Egypt's territorial control during the New Kingdom.

Pyramid of Giza. But there are many historians, geologists, and archeologists who are convinced that these two monumental structures were constructed around the same time. If we place Gobekli Tepe within the chronological timeline and context of Holy Scripture, this archaeological discovery just might be evidence of Seth and his descendants who inscribed their knowledge into monolithic stones so they would not be forgotten. Excavations in process since 1994 have revealed only a fraction of what appears to be a palatial complex with polished limestone floors. Giant pillars, weighing over twenty tons each, display intricate relief carvings of foxes, scorpions, wild boars, birds, lions, insects, human figures, and snakes. These colossal stones align precisely with constellations of the stars as they migrate across the heavens.

Discoveries like this are transforming our perception of human history. The temple of Gobekli Tepe was an enormous accomplishment, one that would be extremely difficult to replicate, even with today's technology. Compounding evidence convinces me to believe our ancestors were far from being the crude, ruddy, cavemen presented to our children in history books today. Radiocarbon dating methods suggest Gobekli Tepe is at least 10,000 years old, but perhaps this archaeological site appears older than the biblical timeline because it was exposed to a CME (Coronal Mass Ejection) or a high intensity solar flare. A flare that would have been ten- to twenty-times more powerful than the greatest ones we have on record today. Perhaps this solar outburst was the force of fire Adam predicted. A solar flare of this intensity would potentially have caused the stone pillars and rocks at Gobekli Tepe to register an older date. Some of the rocks at Gobekli Tepe show signs of "sintering," a geochemical process causing rocks to fuse. A simple example of sintering is how ice cubes stick to each other when they begin to melt. Sintering takes place when rocks are exposed to extreme, high temperatures.

A 2014 newsletter about Gobekli Tepe by *DAI Press* explains that sintering at this location took place after the site was buried with layers of sand and rubble" "Sinter develops under quite particular conditions, in this case only after burial with sediment."[1] Research by physicist Anthony Peratt presents the possibility of sintering by high magnitude, Earth-directed, solar outbursts.[2] I'm thinking Gobekli Tepe may have been intentionally buried to preserve the knowledge of Seth and his descendants as they expected a worldwide, catastrophic event.

Righteous Enoch

The patriarch Enoch from the lineage of Seth was 333 years old when Adam died, and he had first-hand knowledge from Adam about who God was. Enoch—high priest, counselor, and spiritual mediator—was credited with being a "revealer of mysteries." He was appointed by God as scribe of the *Ennead.* In Greek, *ennead* means a collection of nine. Before Noah's flood, nine patriarchal fathers reigned after Adam

---

[1]Laura Dietrich et al., "Investigating the function of Pre-Pottery Neolithic stone troughs from Göbekli Tepe – An integrated approach," *Journal of Archaeological Science: Reports*, 34, Part A (December 2020). https://doi.org/10.1016/j.jasrep.2020.102618.

[2]A L Peratt and W F Yao, "Evidence for an intense solar outburst in prehistory," *IOP Publishing*, (December 2008). https://plasmauniverse .info/downloads-petros/Peratt&YaoAurora-PrehistoryPhys-Scr-T131 ,2008c.pdf.

as the righteous ones: Seth, Enos/Enosh, Cainan/Kenan, Mahalalel, Jared, Enoch, Methuselah, Lamech, and Noah. Enoch was appointed by God to create an authentic record of experiences during this time period, along with preserving the sacred knowledge passed down by Adam. The written word was essential for organizing and leading people in paths of righteousness. It was the French Egyptologist, Schwaller De Lubicz, who discovered glyphs and relics on walls at the Temple of Luxor were incomprehensible unless viewed in conjunction with those appearing on the opposite side, indicating people believed the writings were so powerful that certain hieroglyphs had to be engraved on two separate tablets to protect this wisdom from being misused. As it is recorded in the *Book of Hebrews*, God's Word has the power to divide soul and spirit. We are to call upon the Lord with an open heart in expectation of hearing His voice.

> *TODAY IF YOU HEAR HIS VOICE, DO NOT HARDEN YOUR HEARTS ... For the word of God is living and active and sharper than any two-edged sword and piercing as far as the division of soul and spirit, of both joints and marrow, and able to judge the thoughts and intentions of the heart.* (Hebrews 4:7-12)

Enoch was a scribe appointed by God Almighty to document the history and science of creation. "Numerous scholars and believers assert that the books of Enoch possess the mysteries of the universe. Some believe they contain information about the Big Bang, black holes, dark matter, and dark energy"[1] Enoch was one of the antediluvian (pre-flood) legends who later became known and revered as Thoth by the Egyptians, Hermes by the Greeks, and Mercury by the Romans. The Egyptian carvings record a story very similar to the *Book of Hebrews* saying that Ra (God) sent for Thoth (Enoch) and brought him to Tuat (the levels of heaven). Hebrews 11:5 says:

> *By faith Enoch was taken up so that he did not see death;*
> *AND HE WAS NOT FOUND BECAUSE GOD TOOK*
> *HIM UP; for he obtained the witness that before his being*
> *taken up, he was pleasing to God. And without faith it is*
> *impossible to please Him, for he who comes to God must*
> *believe that He is, and that He is a rewarder of those who seek*
> *Him.* (Hebrews 11:5-6)

The books of Enoch tell of how he traveled through the ten levels of heaven with two angels and became an intercessor for demons begging for God's mercy. Enoch's writings became the repository of the greater Egyptian mysteries, and his work became referenced as The Science of Thoth, who was "master of divine words and sacred writings."[2]

A total of 366 books were written by Enoch. The first verse of the second book of Enoch (Slavonic Enoch) tells us he was

---

[1] Allen Austin, *The Middle of the Earth*, (Xulon Press, 2011), 83.

[2] R. A. Schwaller de Lubicz, *Sacred Science: The King of Pharaonic Theocracy*, (Inner Traditions, 1982), 7.

an eyewitness to the "wise, great, unimaginable, and undeniable realm of Almighty God, to the very wonderful, glorious, bright and many eyed station of the Lord's servants, and to the inaccessible throne of the Lord, and to the degrees and manifestations of the spiritual hosts, and to the indescribable administration of the multitude of the elements, and to the various spirits and inexpressible singing of the host of Cherubim, and to the limitless light" (2 Enoch 1:1). Only thirty-two years passed from the time of Adam's death to the time of Enoch's ascension as is recorded by the Second Book of Enoch.

The original discovery in 1946 of thousands of cuneiform inscriptions and scrolls within the Qumran Caves breathed new life into a more complete and authentic understanding of human history. This was just two years before Israel became a nation in 1948. Much later, in the 1960s, ancient cuneiform tablets referencing the writings of Enoch were found alongside the Dead Sea Scrolls. This discovery led researchers to investigate and confirm the authenticity of more than seventy books written by the pre-flood patriarch Enoch. About thirty years later, in 2003, over fifty scholars and linguists from around the world met at the Second Enoch Seminar to compare research and interpretations of the tablets. The consensus of thought by those at the seminar was that the ancient scribes—independent of social and cultural differences—had transcribed Enoch's books not only to preserve history but to instruct people and lead them along paths of righteousness. They determined these records of antiquity were preserved by eloquent and conscientious historians and authors who were capable of very intricate record keeping.

Biblical scholar Michael Anthony Knibb refers to the study of 1 Enoch as a Jewish work that is "known to us only because it was preserved and transmitted by Christians; and how best to account for the formation of the book as it is known to us in its most complete form, that is in the Ethiopic version."[1]

Zion & The Pillar of Enoch

Enoch was the father of Methuselah. When he was taken up by God at 365 years old, his son "Methosalam (a.k.a. Methuselah), and his brethren, all the sons of Enoch, made haste, and erected an altar at that place called Achuzan, whence and where Enoch had been taken up to heaven."[2] Historian and Rabbi Louis Ginzberg discovered scroll fragments recording that Achuzan was a sacred location in the middle of the Earth where Adam was created.[3] The Syriac "Cave of Treasures" and 2 Enoch 71:35 reveal Adam's burial was also at this location. Melchizedek was "a priest at Adam's final burial place in the middle of the Earth."[4]

---

[1]Michael A. Knibb, *Essays on the Book of Enoch and Other Early Jewish Texts and Traditions*, (Brill Academic Pub, 2008), 2.

[2]2 Enoch Ch. 68 v.6.

[3]Louis Ginzberg, *The Legends of the Jews: From the Creation to Exodus: Notes for Volumes 1 and 2 (Volume 5)*, (Baltimore and London: The Johns Hopkins University Press, 1925, 1953).

[4]Orlov, Boccaccini, Zurawski, *Cave of Treasures*, 46.

"O great and mighty God, The LORD of hosts is His name; great in counsel and mighty indeed . . . who has set signs and wonders in the land of Egypt" (Jeremiah 32:18-20).

As was pointed out earlier, John Taylor, Piazzi Smyth, Schwaller de Lubicz, Allen Austin, and others collectively agree the Great Pyramid is the original Mount Zion at the middle of the Earth. And this is where Enoch was walking with God when he was taken up to be with Him. Blow the trumpet in Zion, sound the alarm! This is where, in the beginning of time, the first man Adam was created. A place where Holy high priests offered prayers and praises to God. It later became known as the "Pillar of Enoch" or "Pillar of Thoth" to local Egyptians, which is one and the same with the Great Pyramid of Giza. The significance of Enoch being at this location when he was taken up to be with the Lord is truly remarkable.

1. It is a confirmation of God's promise to redeem those who choose to walk with him as Enoch did.
2. It is a reminder of the message that Adam faithfully conveyed for more than nine hundred years about God's promised gift of eternal life.
3. It is a foreshadowing of God's promise to return for His bride who will be caught up to be with Him before the tribulation.

Flinders Petrie, a British Archaeologist and Egyptologist born on June 3, 1853, calculated that the Great Pyramid of Giza is situated exactly at the center of Earth. In 1923, he was knighted for his services to British archaeology and Egyptology.[1]

In the course of his questioning he (Herodotus) encountered one Manetho, an Egyptian High Priest, scholar, and Historian, with whom he conversed at length through the agency of an interpreter. Manetho informed his distinguished guest that the architect of the huge mass of stone was one "Philition," or "Suphis," of a people known as the "Hyksos," that is "Shepherd Kings." According to Manetho, the Shepherd Kings were "a people of ignoble race" who came from some unknown land in the East; they were a nomadic band who numbered not less than 280,000 souls; they brought with them their families and all mobile possessions, including vast flocks of sheep and herds of cattle; and they "had the confidence to invade Egypt, and subdued it without a battle." This same people, said Manetho, overthrew the then-reigning Dynasty, stamped out idolatry and endeavored to firmly establish in the place thereof the worship of the One true God having completed the Great pyramid, migrated eastward into the land afterwards known as Judea and founded there the city of Salem, which later became Jerusalem, the Holy city (Herodotus 12).[2]

However, construction of the Great Pyramid was never completed. The final capstone was never mounted at the top.

---

[1]"Flinders Petrie archaeologist," https://www.age-of-the-sage.org/archaeology/flinders_petrie.html.

[2]Allen Austin, *The Middle of the Earth*, (Xulon Press, 2011), 135.

No one knows for sure if it was always without a capstone, but accounts of visitors even during the time of Christ report it did not have a capstone. According to prophetic Scripture, this capstone was the crowning cornerstone of perfection, but the builders refused to accept it (Mark 12:10). Is it possible construction of the great pyramid was never completed because God's time is not our time? Isaiah 4:7 describes a prophetic day when the truth is revealed despite all the hindrances to its completion, and the final capstone will be brought forth with shouts of joy crying, "Grace, Grace unto it."

Perhaps the significance of the Great Pyramid of Giza and crowning capstone have been kept behind a veil of deception for thousands of years, awaiting fulfillment of another prophecy recorded by King David. "The stone which the builders refused is become the head cornerstone" (Psalm 118:22). Behold, I tell you a mystery. The last generation will not pass away. Praise be to our Lord God Almighty! For He is the Alpha and the Omega and has dominion overall, and His Truth will be revealed from beginning to end. Our present time parallels ancient history, and this tangled web of deception that has prevailed for millennia is being unscrambled. Yeshua (Jesus) said in His Olivet discourse, "As in the days of Noah, so shall also the coming of the Son of man be" (Matthew 24:37). We are living in a time of great deception, and those seeking the Truth will be set free.

CHAPTER 6

NEPHILIM WERE ON THE EARTH IN THOSE DAYS

A short time before Enoch was taken up to be with the Lord, he instructed and encouraged people to remain holy and separate from those who were rebellious, rowdy, and licentious. He warned the righteous line of Seth about a great deception that would take place with deviant angels conspiring to descend upon the Earth: "There were giants in the Earth in those days and also after that, when the sons of God came in unto the daughters of men, and they bore children to them, the same became mighty men which were of old, men of renown" (Genesis 6:5). The supernatural influence of the fallen angels produced genetic traits in their offspring resulting in characteristics like those of Goliath of Gath. The one whom David conquered with the help of his Father God, and Og of Bashan whose bed is described In Deuteronomy 3:11 as being nine cubits long and four cubits wide. Author Gary Wayne points out that humans mating with humans could not possibly have produced "a new, physically superior race of humanoids of that sort of magnitude, unless the term giant is greatly overstated, for Nephilim were thought to be a powerful race of giants."[1]

God gave Moses a condensed version of events during his time in Egypt, which he articulately recorded in the book of Genesis: "Moses wrote that some of the angels of God left their prior estate and came down to Earth and mated with women, which produced a race of super-human hybrids known as the Nephilim."[2] Yes, there were hybrids on the Earth during this time. Why do you think Egyptian monuments, Greek mythology, and Indian art display animals with human attributes and vice versa? I wonder if our current biological studies with GMOs, cloning, gene splicing, and DNA mapping will result in something similar to what took place in the days of Noah, as Holy Scripture says, today we live in parallel times. Finding the "Truth" about our history requires courage and keen discernment to navigate conflated and convoluted stories about human history and who God is. Since the beginning of creation things went terribly wrong. But those of faith stood firm in their belief, trusting that Almighty God would provide the confidence and strength they needed for whatever situation that arose. As a result, they became empowered leaders,

---

[1]Gary Wayne, *The Genesis Six Conspiracy*, (Trusted Books, 2014), 6.

[2]Rob Skiba, *Genesis and the Synchronized, Biblically Endorsed, Extra-Biblical Texts*, (King's Gate Media, 2013), 1.

strengthened by their Heavenly Father, who brought them through fiery trials[1]

After righteous Enoch's ascension, a boom of deception dropped as the giants matured and began using supernatural, angelic powers to dominate and control people. The record of Enoch describes these cross breeds as having long, narrow faces, prominent cheekbones, elongated jaws, thin lips, and slanted eyes. Artifacts from the Ubaidian and Vinca cultures demonstrate these distinct features. The ancient Mayan, Easter Island, Sumerian, and Mohenjo Daro artifacts of Pakistan exhibit similar features. Professor Gabriele Boccaccini, founding director of the Enoch Seminar, explains that the cosmic rebellion of fallen angels was the mother of all sins.

The original sin which corrupted and contaminated God's creation and from which evil relentlessly continues to spring

---

[1]https://commons.wikimedia.org/wiki/File:Lizard-headed_nude_woman_nursing_a_child,_from_Ur,_Iraq,_c._4000_BCE._Iraq_Museum.jpg
Description
English: Terracotta figurine of the co-called "lizard-shaped" face. The woman is nude and is breastfeeding an infant. The hair piece was added and is made of bitumen. Height 14 cm. From Ur, Iraq. Late Ubaid period, c. 4000 BCE. On display a the Iraq Museum in Baghdad, Iraq.
Date
14 March 2019, 09:31:19
Source
Own work
Author
Osama Shukir Muhammed Amin FRCP(Glasg)

forth and spread. As God said to the angel Raphael: "The whole Earth has been corrupted through the works that were taught by Azazel; to him ascribe all sin!" (1 En 6:8). In a cosmic battle the rebellious angels were defeated by the good angels and imprisoned in chains "in a hole in the desert which is in Dudael . . . (until) the day of the great judgment" (1 En 6:4-6). The giants, the monstrous offspring of the unnatural union between angels and women, were killed (10:9-10), but their immortal souls survived as the evil spirits—and continue to roam about the Earth (15:8-10). As disturbing as this idea can be, God's reaction limited but did not eradicate evil, until God will put an end to this evil world and will create a new qualitatively different from and discontinuous with, what was before.

The New Testament warns women about protecting themselves from the influence of fallen angels, by having a seal, a sign, or a symbol of authority upon their head (1 Corinthians 11:9-10). Jude 1:6-7 confirms the significance of angels mating with humans. Men lusting after angels whom Lot invited to the safety of his home were threatening to break in. It was before the flood of Noah, that sacred knowledge of the patriarchal fathers was usurped and a giant race of crossbreeds began ruling the world with supernatural influences. Wisdom of the heavenly sciences was hijacked for sorcery and deception. Generations following Enoch began to forget about the Lord God Almighty. They became mesmerized by chants, trinkets, and powers of the cunning Nephilim, who lured people into thinking they could be god. The Hebrew word Nephilim means "those who have fallen." For six hundred years after Enoch ascended, Earth's population grew with these greedy, giant-sized hybrids whose insatiable desires led to violence, tyranny, and human sacrifice. It was a frontal attack to destroy the "seed" of promise made by God from the beginning of creation. Narrative accounts of the gods became abstract, and future epochs would merge history with legend. Amazing discoveries of giant human bones were destroyed. Their existence was completely discredited and stories of bestiality became myth. The depravity of mankind prior to the great flood of Noah was absolutely terrifying. Is it any wonder that God should intervene with a judgment so disastrous as the great flood of Noah?

"But Noah was a Righteous Man"

Inscribed cuneiform tablets recording the story of the flood indicate the city of Shurappak was the home of Utnapishtim, the legendary Babylonian name for Noah. The Bible tells us Noah was a righteous man who walked with God. According to Genesis 5:32, sometime after Noah was five hundred years old, he became the father of Shem, Ham, and Japheth. Then in Genesis 7:11, we are told it was in the six hundredth year of Noah's life when all the fountains of the great deep burst forth. So how many years did it take for Noah to build the ark? Now according to Genesis 11:10, Shem became a father to his first child at one hundred years of age, just two years after the flood. This indicates Shem entered the ark when he was around ninety-seven years old. He had a wife, but their first child was not conceived until two years after the flood. It is fascinating to think about the lives these people lived. "By faith Noah, when warned about the things not yet seen, in holy fear built an ark to save his family. By his faith he condemned the world and became heir of the righteousness that is in keeping with faith" (Hebrews 11:7). We can only speculate on the number of years it took to build the ark. I suspect it took Noah seven years to build the ark, but who knows, it may have taken seventy. God gave him detailed instructions. "Make thee an ark of gopher wood; rooms shalt thou make in the ark, and shalt pitch it within and without with pitch. [. . .] The length of the ark shall be three hundred cubits, the breadth of it fifty cubits, and the height of it thirty cubits" (Genesis 6:14-16). These measurements reveal that the ark was approximately the size of a three-story apartment building, with over 100,000 square feet of living space for the animals alone.[1]

---

[1] https://commons.wikimedia.org/wiki/File:Ark_structure.jpg
Description
English: Durupınar – Alleged landing site of Noah's Ark near Dogubayazit, Turkey, 17 miles south of Mount Ararat. Courtesy Dr. Lorence G. Collins.
Date
13 October 2009
Source
Own work
Author
Rudolf Pohl

Ark Protruding Mt Judi

In 1985, ship builder David Fasold, archaeologist Ron Wyatt, and geophysicist John Baumgardner set out to research a protruding formation near Mount Judi, what once was known as Cudi Dagi south of Lake Van in Turkey. It was a boat-shaped mound named after Turkish Army Captain Ilhan Durupinar, who identified the formation while looking at a Turkish Air Force aerial photo during a mapping mission for NATO in 1959.

The three men trekked the mountainous territory of the Kurds with several instruments. One of them was a sensitive radar device capable of detecting mineral content and formations to a depth of forty feet below the surface. A subsurface interface radar (SIR) machine would be used to detect beams and cross beams. A molecular frequency generator would be used to measure and map a sequence of evenly spaced iron concentrations. The length and breadth of the shipwreck-looking protrusion aligned with measurements recorded in the Bible as instructions for Noah to build the ark.

Fasold's experience as a merchant marine officer and salvage expert provided him with unique insights as to how the ark was constructed. His knowledge about the history of shipbuilding, along with records of Holy Scripture, led him to think massive bundles of reeds were covered with a resinous tar in order to shape and form the walls of the ark. It would have been a material similar to what was used to cover the bulrushes of the ark carrying baby Moses in a stream to the Pharaoh's daughter.[1] About 5,000 years ago Akkadian shipbuilders used a mixture known as KPR, a bituminous brimstone resin blended with pumice and natron. Fasold points out that natron would have been a catalyst for the hardening process.[2] "The way ancient Egyptians made a cement mortar that has lasted thousands of years was recently deduced by a French chemist, Joseph Davidovits . . . Specifically the ancient Egyptians mixed lime and natron."[3] The great ship of Noah's ark was coated inside and out with this pitch, forming a virtual concrete barrier of significant strength.

---

[1] David Fasold, *The Ark of Noah*, (Wynwood Pr., 1989), 267.

[2] David Fasold, *The Ark of Noah*, (Wynwood Pr., 1989), 271.

[3] http://www.housebarracom/EP/ep15/22cement.html.

Constructing Noah's Ark

Much to Fasold's surprise, measurements of cross beams and locations of iron concentrations revealed "a large rectangular opening in the hull 201- feet by approximately 26- feet, maybe slightly more. What was this open area? A ramp or a corridor? Certainly, an enormous space to not contain transverse bracing!"[1] This information led David Fasold to think the ship had a moon pool, one that would facilitate drainage and aid ventilation. As water rose into this large opening in the bottom of the ark, air would have been forced to circulate throughout the entire ship. The moon pool would also have provided access to fish and helped with stabilization of the boat. Wikipedia

---

[1]David Fasold, *The Ark of Noah*, (Wynwood Pr., 1989), 117.

describes a moon pool as "an opening in the floor or base of the hull, platform, or chamber giving access to the water below, allowing technicians or researchers to lower tools and instruments into the sea. It provides shelter and protection so that even if the ship is in high seas or surrounded by ice, researchers can work in comfort rather than on a deck exposed to the elements. A moon pool also allows divers, diving bells, ROVs, or small submersible craft to enter or leave the water easily and in a more protected environment."[1]

Flavius Josephus, c. 90 AD, the famous Jewish historian, quoted Berosus the Chaldean, c. 290 BC, saying that in his days the remains of Noah's great ship were shown to visitors by locals in the vicinity, and tourists took home pieces of the ark as good-luck charms. "It is said there is still some part of this ship in Armenia, at the mountain of the Cordyaeans; and that some people carry off pieces of the bitumen, which they take away, and use chiefly as amulets for the averting of mischiefs." This recorded documentation indicates the location of the ark was not inaccessible prior to the birth of Christ. At some point the ark was covered by mud and a lava flow which caused future generations to lose its location until an earthquake exposed it again at a later date as the protruding outline of the ark at Al-Judi.

---

[1] "Moon pool," https://en.m.wikipedia.org/wiki/Moon_pool.

The Stone Things

Today we know of more than five hundred flood stories from every tribe and culture worldwide dating back to the dawn of writing. Noah's ark and the great flood was a historic event not only documented by biblical records but by several artifacts exhibited in museums today. These precious relics are a window to the truth about creation and the flood. For example, "The Eridu Genesis" is an inscribed tablet from 2300 BCE containing one of the oldest cuneiform texts from Mesopotamia about creation and the flood of Noah. Atrahasis is an eighteenth century BCE Akkadian epic about an exceedingly wise man who was warned about an epic deluge. The name, Atra-Hasis, is listed among the Sumerian kings of Shuruppak on the King List. Stories handed down by oral tradition from generation to generation were embellished, and at times contorted, and yet they are a significant confirmation of the biblical account. The Epic of Gilgamesh is another engraved tablet about the flood referring to the weights that were used to balance the great ship built by Noah. "You can't cross the waters without the stone things." There's a fascinating YouTube video showing how a ship modeled after Noah's ark managed to avoid being capsized by enormous waves with "the stone things" beneath it, https://www.youtube.com/watch?v=1O8wGjwyS7o.[1]

---

[1] AnchorStones, "Tests Show Drogue Stones Effectiveness on Noah's Ark," *Youtube*, https://www.youtube.com/watch?v=1O8wGjwyS7o.

There is a village named Arzap—meaning "The Place of Eight"—about fifteen miles from the protruding outline of Noah's ark on Mount Judi. This is where several massive anchor stones are located. They are similar to much smaller ones that were discovered closer to sea level, adjacent to old shipwrecks. The stones at Arzap are much larger, weighing from four to ten tons each; they are thousands of miles away from the nearest ocean and over 3,000 feet above sea level. The unbroken ones have smooth, tapered holes drilled in an upward angle through the top section. This design would allow a thick rope to slip through a larger opening on one side of the hole and be knotted tightly to fit snugly into the narrow side, preventing the hole in this giant stone from being fractured if lifted out of the water.[1]

---

[1]https://commons.wikimedia.org/wiki/File:DrogueStoneFasold.jpg
Description
English: David Fasold, a promoter of the Durupınar site, stands with a Drogue Stone in Arzap ((Kazan) Türkiye) (crosses are believed to have been added later)
Date
25 February 2007 (original upload date)
Source
en.wikipedia. Transferred to Commons by User:Tuckerresearch using CommonsHelper.
Original source: Photograph created by Robert C. Michelson using an Asahi Pentax 35mm SLR camera with Kodak Kodachrome 200/24° film with permission of David Fasold (subject)
Author
Robert C. Michelson Robert C. Michelson on en.wikipedia in conjunction with the SEPDAC Corporation.

Drogue Stone Fasold

David Fasold standing next to a 6-foot-tall Drogue Stone in Arzap (Kazan) Türkiye).

These stately objects were recognized by Ron Wyatt in the late 1970s and identified by Fasold as drogue stones for stabilizing a large ship. They would prevent a boat from being capsized if suddenly struck by a surprise wave and in general provide for a smoother ride. These drogue stones would also serve to keep the boat directed toward oncoming waves. Early Christians who recognized them as relics from the biblical story of Noah's flood are believed to have carved the crosses seen on them today. Similar but smaller anchor stones were later discovered by the Mediterranean Sea, off the coastlines of Spain, France, Ireland, and even California.

Fasold was forced in an Australian court of law to repudiate his findings when a geologist and critic of creation, Ian Plimer, brought a suit against him. But his biographer friend June Dawes records that on his deathbed, David told her he was convinced by his research and findings that the anomaly adjacent to Mt. Cudi was truly the ark of Noah (June Dawes 2000, pg. 184). I thank God for the inspirational insights brought to light by this courageous and talented shipbuilder named David Fasold. "The righteousness of the blameless will smooth his way, But the wicked will fall by his own wickedness" (Proverbs 11:5).

Scriptures tell us Noah was six hundred years old and his son Shem was ninety-eight when they boarded the ark. Not only was Noah warned of a flood, but God told him, "The end of all flesh has come before Me; for the Earth is filled with violence because of them; and behold, I am about to destroy them with the Earth" (Genesis 6:13). Notice that God does not say He will 'drown' all life but rather, He was going to blot out every living thing within the Earth. "And the Lord said, 'I will blot out man whom I have created from the face of the land, from man to animals to creeping things and to birds of the sky, for I am sorry that I have made them; and behold, I am about to destroy them with the Earth'" (Genesis 6:7).

CHAPTER 7

DYNAMIC SYNERGY
AND CHAIN REACTIONS

After much searching and pondering the Earth's features as outlined by Holy Scripture, another question remained. What activation energy triggered the explosion of all the fountains of the great deep? David, the King of Israel, wrote about God's awesome power and how He uses natural forces to carry out His will. Following is a comprehensive study of scientific evidence in support of Holy Scripture recording the masterfully orchestrated event of Noah's flood.

God's Word suggests the activation energy for this global calamity may have been an extraordinary solar flare with this verse, "Who makes His angels winds, his ministers a flame of fire" (Hebrews 1:7). My research has convinced me that after construction of the ark was completed, a time of persistent and recurrent solar flares began. They caused the night skies to light up, and people watched in awe as prismatic hues danced beneath the sphere of the vaulted dome. Animals spooked by the radical changes in their environment looked for a place of safety. After many years of construction in the tropical, temperate, environment of pre-flood Earth, Noah's amazing ship must have been surrounded with lush vegetation and perhaps a moss-covered ramp led up to a cave-like opening within the ark. Animals from near and far instinctively sought refuge and migrated to this mysterious entrance deep within the jungle. Noah and his family were instructed by God to select male and female pairs from each species. As the youngest animals were chosen, they scurried into the clandestine opening and nestled down within the safety of the great ship's hay-lined stalls.

HAVE YOU EVER IN YOUR LIFE COMMANDED THE MORNING AND CAUSED THE DAWN TO KNOW ITS PLACE; THAT IT MIGHT TAKE HOLD OF THE ENDS OF THE EARTH, AND THE WICKED BE SHAKEN OUT OF IT? JOB 38:12-13

Solar Flare

With the doors of the ark secured, Noah led his family in prayer as the Earth beneath their ship began to tremble. Another solar flare erupted, but this time it was so immense that time-dependent penetration of the sun's magnetic forces no longer had a delayed impact. A great geomagnetic storm began penetrating to the core of Planet Earth with an unprecedented, monumental, solar wind. One with a magnitude so great the power generated by its solar particles became trapped within the Earth's magnetic field, and enhanced by the vaulted dome.

Volumes of scientific research have been collected with regards to the effects of solar flares upon our planet. Only in the recent past have scientists had access to high-tech information gathered by satellite systems tracking and documenting changes within Earth's magnetic field when it is impacted by solar radiation. In light of the following research, allow me to present for your consideration the possibility of an immense solar flare creating such a great disturbance within Earth's molten core that it caused all the fountains of the great deep to burst forth as recorded in Genesis 7:11.

Electric currents generated within Earth's molten core are known to produce a magnetic field at the Earth's surface. Scientific research confirms trapping of charged solar particles within the Earth's electric/magnetic field produces tension at the core/mantle boundary. This discovery was made by scientist Casenave Anderson as he observed the aurora borealis dancing with greater brilliance in the night sky after a huge solar flare had taken place. He researched and recorded scientific details about what had caused the northern lights to be so extravagant. His investigation led him to realize this phenomenon was related to a solar flare because the Earth's mantle is "an electrical conductor [. . .] and the electrical field induces currents in the lower mantle which interact with Earth's primary magnetic field through the Lorentz force, producing a torque on the mantle (see Rochester 1960)."[1] This magnetic tension reveals a relationship between the Earth and the sun. After a solar flare takes place, the northern lights become exaggerated and present a fascinating display as the aurora borealis lights up the night sky. High-speed, electrically charged particles emitted from the sun at the time of a solar flare also generate external currents known as a solar wind. "The solar wind is responsible, directly or indirectly, for most of the geomagnetic and energetic particle phenomena in the magnetosphere."[2]

I felt led to contemplate this scientific information by taking into consideration how a solar flare would have impacted the vaulted dome. I wondered if the dome's composition would have caused a diffraction of high energy particles contributing to the formation of a luminous arc between the north and south poles. This concept was inspired by Job 38:12-13, which suggests a powerful circuit of electrical charges were conducted between the north and south poles as God took hold of the ends of the Earth: "Have you ever

---

[1] Anny Cazenave, *Space Geodesy & Geodynamics*, (London: Academic Press Inc., 1986), 309.

[2] Billy McCormac, *Earth's Particles & Fields*, (New York: Reinhold Book Corporation, 1968), 455.

in your life commanded the dawn to know its place; That it might take hold of the ends of the Earth, And the wicked be shaken out of it?"

This led me to thinking the vaulted dome must have played an important role in directing the energy of an intense solar flare to Earth's north and south poles. The prophetic words saying He commanded the dawn to know its place indicate God's power to rule the sun and use it for His purposes. Also, I could not help but think, "What better way could there be to describe the ends of the Earth than with the north and south poles?"

These revelations by Job launched my fascination to discover scientific evidence for a shift of Earth's polar axis. It also led me to question and pursue more information about the immensity of an electromagnetic storm so great it had enough power to trigger an explosion from the great deep. And perhaps there was evidence for a solar flare agitating the molten core of Earth to such an extent that alternating portions of its foundation began to heat up to what is known as the "Curie point."

It was observable evidence at Earth's surface leading me to understand how this exceptional event may have taken place. Scientific research and reliance upon the framework of Holy Scripture led me to discover that high-speed electrically charged particles from the sun do cause Earth's molten core to flare. According to geophysicist Martin Walt, trapped radiation plays "... an important role in many geophysical processes" and the motion "... of trapped ions and electrons produces a magnetic field opposing the field of the Earth's core. In many ways, some only vaguely understood, the trapped population acts as a coupling agent transferring energy, momentum and mass between interplanetary medium and the earth's atmosphere. This transfer is many-faceted."[1]

This information substantiated the idea that a huge solar flare was the activation energy causing all the fountains of the great deep to burst forth (Genesis 7:11). As I dug deeper, it became evident portions of the pillar-like foundation began to melt as Earth's molten core flared, but God had created our planet to endure such a catastrophic event. The melting pillars surrounding the subterranean chambers released minerals into the water flowing past them as it heated up and began turning into steam. Question is, did all the fountains of the great deep have something to do with the transformation of fresh water into the saltwater of our oceans today?

Psalm 18:14 presents us with further details about the dynamics of this event: "And He sent out His arrows, and scattered them, And lightning, flashes in abundance, and routed them." Well, it just so happens solar flares play a significant role in the generation of lightning according to the following article named the "Sun's Magnetic Field Triggers Lightning Strikes on Earth."

As lightning flashes were routed, pressure continued building within the labyrinth of underground caves. As the expansion of steam took place within a confined space, it

---

[1] Martin Walt p 4-5

would have facilitated the absorption of minerals and elements being released by the surrounding melting pillars of basalt. As these minerals and elements in a vaporous solution flowed through the underground chambers, a pipe-like conduit formed for the transmission of electrical/magnetic disturbances being generated by the massive solar flare. It was a dynamic situation, one that may have been further enhanced by a luminous arc beneath the vaulted dome conducting currents between the north and south poles. "Strong magnetic disturbances—cause electrical currents to flow through pipes, transmission wires and other conductors, leading to corrosion."[1] Electrical charges conducted through these channels of steaming hot water trapped beneath the surface of Earth must have resulted in the corrosion and release of minerals into the water.

Holy Scripture tells us a cleft channel provided a passageway. Perhaps this verse is presenting confounding proof for the dynamics of steaming hot water flowing through a labyrinth of caves and becoming the mechanism for an explosion of all the fountains of the great deep during the flood of Noah. For Job 38:25 tells us, "Who has cleft a channel for the flood," and the second part of this verse explicitly suggests the cleft channel became a conduit for the transmission of electrical charges because this verse continues to point out the cleft channel provided "a way for the thunderbolt." In simplistic terms, this verse is describing the dynamic transmission of energy from a solar

---

[1]Candace Savage, *Aurora*, (Vancouver: Greystone Books, 1994), 130.

flare through a cleft channel that was "a way for the thunderbolt" (Job 38:25).

As pressure increased and chemical reactions took place within the underground chambers, it was the sequential effects of a great solar flare that excited conditions to such an extent it triggered the release of a hydrothermal, supersonic explosion. One so extreme it caused Earth's polar axis to shift.

Then the Earth shook and quaked; And the foundations of the mountains [the roots of the mountains still beneath the surface] *were trembling. And were shaken, because he was angry. Smoke went up out of His nostrils, And fire from His*

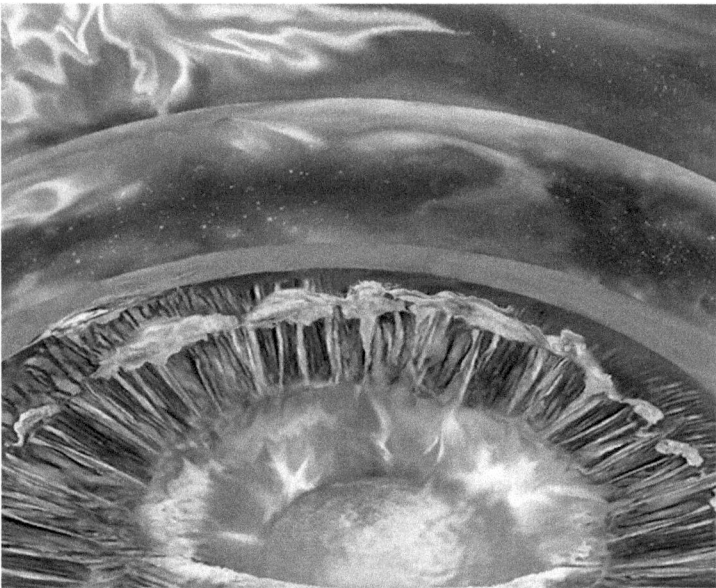

Pre Explosion

mouth devoured; Coals were kindled by it. He bowed the heavens also, and came down with thick darkness under His feet. . . . And He sent out His arrows, and scattered them. And lightning flashes in abundance, and routed them. Then the channels of water appeared, And the foundations of the world were laid bare At Thy rebuke, O LORD, At the blast of the breath of Thy nostrils (Psalm 18:7-15).

Explosions from the Great Deep

"In the six hundredth year of Noah's life, in the second month, on the seventeenth day of the month, on the same day all the

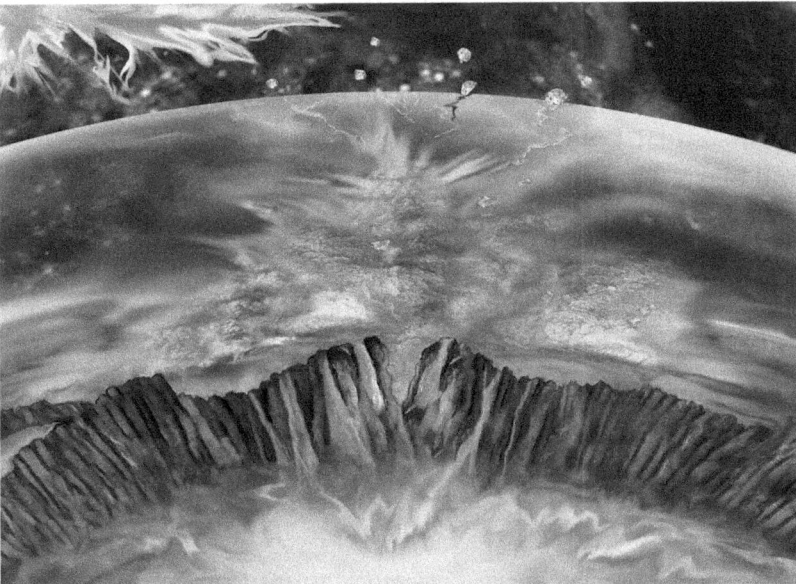

Explosion from the Great Deep

fountains of the great deep burst forth, and the floodgates of the sky were opened" (Genesis 7:11). The exponential thrust of this explosion was similar to, but much greater than, the nuee ardente that took place in 1980 at Mt. St. Helen. The explosion of steam in beautiful Skamania, Washington, was minor in comparison with what took place at the onset of Noah's flood. The steam exploding from storehouses of the great deep 4,356 years ago was exponentially greater than this. It resulted in the ballistic discharge of Earth's crust above ground zero.

Prior to this event there were no ocean basins. All the fountains of the great deep bursting forth was a blast that defied the limits of both lithostatic pressure and gravity. Its supersonic velocity resulted in removal of Earth's crust, leaving the pillar-like foundation of Earth exposed at the ocean floor, as is recorded by Psalm 18:15: "Then the channels of water appeared, and the foundation of Earth was laid bare." The vertical thrust of this explosion shattered the overlying crust and blasted it into the protective barrier of the vaulted dome.

As caprock above ground zero impacted the vaulted dome, great chunks of it were ballistically ejected and propelled into outer space. These chunks of ice go around the sun on a highly elliptical orbit as comets and are often referred to as "dirty snowballs" because they are made of ice, dust, and small, rocky particles. "Comets are cosmic snowballs of frozen gases, rock, and dust that orbit the Sun."[1] If comets

---

[1]"Comets," *Solar System Exploration,* https://solarsystem.nasa.gov /asteroids-comets-and-meteors/comets/overview/?page=0&per_page

contain water in the form of ice, and Earth is the only planet in our solar system containing water, my hypothesis about them originating from the vaulted dome seems rather plausible. And most recently it was discovered comets contain glycine, an organic compound of protein, which is an amino acid essential to and evidence for life on Earth. This discovery also substantiates the concept of comets originating from our planet.[1]

According to the Ryrie Study Bible, 1656 years after creation week marks the six-hundredth year of Noah's life according to the genealogical succession recorded by Genesis chapters five and six. This was the epic year when all the great deep subterranean storehouses burst open in a simultaneous sequence of explosions, resulting in the geologic transformation of Earth. The vaulted dome once supported by the dipolar magnetic poles of Earth was completely destroyed by this explosion. An illuminating rendition of this momentous event was preserved and passed down by ancient Chinese storytellers: "Reading from Huai-nan Tzu, the cause of the cosmic disaster is said to be that 'the four poles collapsed,' that is, the Earthly supports holding up the sky gave way."[2] Sure sounds to me like this describes a collapse of the vaulted

---

=40&order=name+asc&search=&condition_1=102%3Aparent_id&condition_2=comet%3Abody_type%3Ailike.

[1]"Comet contains glycine, key part of recipe for life," *Phys Org*, (May 2016), https://phys.org/news/2016-05-comet-glycine-key-recipe-life.html.

[2]Anne Birrell, *Chinese Mythology: An Introduction*, (Baltimore and London: Johns Hopkins University Press, 1999), 69.

dome. Just imagine the impact of this grand event. Not only did it cause enough rain to cover Earth's still formless surface, but it resulted in post-flood conditions that had an enduring influence upon the global climate. The explosion of Mt. St. Helen was modest in comparison, and yet, it had a dramatic impact upon Earth's climate for a short period of time.

The Illusive Anomaly—A Shift of Earth's Polar Axis

According to Bruce Heezen and Maurice Ewing, the paralleling crest of the Mid-Ocean Rift displays as a general pattern, a large positive magnetic anomaly associated with the entire rift system.[1] This geologic fact suggests the 43,000-mile-long rift valley was heated to the Curie point and lost its original magnetization all at the same time. It seems to me this is outstanding evidence for a geologic event of great magnitude.

The term "positive magnetic anomaly" means the current magnetic line of force at Earth's surface is the same as, or concurrent with, the magnetic field of iron-bearing particles within a body of rock.[2]

Paleomagnetism is the remnant magnetism in ancient rocks that records the direction and strength of the Earth's

---

[1]Tharp and Frankel, *Mappers of the Deep; Natural History*, 62.
[2]Peter J. Smith, *Earth Is A Magnet*, (New York: Trewin Copplestone Books Ltd., 1981).

magnetic field at the time of their formation. When a magma cools through the Curie point, its iron bearing minerals gain their magnetization and align themselves with the Earth's magnetic field recording both its direction and strength. As long as the rock is not subsequently heated above the Curie point it will preserve that magnetism. However, if the rock is heated above the Curie point, the original magnetism is lost, and when the rock subsequently cools, the iron-bearing minerals will align with the current magnetic field.[1]

The paleomagnetic record is presented as supporting evidence for the ambitiously promoted theory of sea-floor spreading. It is proposed by modern theories that the paleomagnetic record presents evidence the central rift is the newest formation created by magma upwelling at the ocean floor and that pillars adjacent to the mid-ocean rift valley were formed after the last polar shift. This geologic phenomenon, known as the "Paleomagnetic Record," alternates across the entire ocean floor like the stripes of a zebra. This pattern suggests every other segment of Earth's foundation got hot enough to record a shift of Earth's axis. Question is, was this a one-time, world-wide event, or were the pillars of basalt at the ocean floor heated to the Curie point multiple times over millions of years?

The theory of Paleomagnetism becomes even more confusing with the introduction of magnetic reversals and

---

[1]James S. Monroe and Reed Wicander, *Physical Geology – Exploring The Earth*, (New York: West Publishing Co., 1992), 310.

magnetic excursions. "When these magnetic reversals occur, the Earth's magnetic polarity is reversed so that the north arrow on a compass would point south rather than north."[1] I questioned if it would even be possible for our planet to sustain the stress of its polar axis repeatedly flipping over with multiple north to south reversals, especially in light of Holy Scripture's assertion the Earth is firmly established. He established the Earth upon its foundations, so that it will not totter forever and ever" (Psalm 104:5). As with other puzzling predicaments in geology, a series of events within extended and incomprehensible time frames are favored for the interpretation of data. No wonder Holy Scripture repeatedly mentions the Earth is firmly established!

Looking at this data through the lens of Holy Scripture suggests that the Paleomagnetic record may be revealing a one-time shift of Earth's axis that took place when all fountains of the great deep burst forth. A one-time extraordinary event that took place when Earth's molten core flared to heat Earth's pillar-like foundation. Job 38:8-11 tells us that immediately following an explosion from the great deep, God placed a bolt and doors:

> *Or who enclosed the sea with doors, When bursting forth it went out from the womb; When I made a cloud its garment, And thick darkness its swaddling band, And I placed boundaries on it, And I set a bolt and doors, And I said thus far*

---

[1] iBid.

you shall come, but no farther; And here shall your proud waves stop?

The concept of "a bolt" triggered the thought of magnetism in my mind because traditionally a bolt is made of iron. Reading this verse of Holy Scripture within context it became apparent the word "bolt" in Job 38:10 seemed to be a description of the iron-bearing particles locking in to record a new magnetic alignment of Earth's polar axis. Foundational pillars containing a greater volume of silica did not heat up to the Curie point, so iron-bearing particles in these pillars maintained their original alignment with true north. Earths current axial rotation is 23.5 degrees from the true north. Job 38:8-11 clearly states that the iron bolt was set in place after the fountains of the great deep had burst forth. Once again, I cannot help but think our Creator intended for this mystery to be revealed during the last days when the times we are living are just like those *"as in the days of Noah"* (Matthew 24:37)?

Does not the Word of God inspire consideration of how precise He is? Luke 19:40 tells us that *"even the stones will cry out."* Austin Gentry explains it well:

Creation will "speak" to what is true—whether we, as humans, acknowledge it or not. When God made the universe, He created it with a certain design. It works a certain way. There is a certain "grain" to reality. There are governing, built-in truths that guide, frame, and endow it with a sense of meaning and flourishing. Scientifically, mathematically, biologically, morally, sexually, relationally, etc. And we can choose to live in a way that "goes with the grain" of reality—or not. We can acknowledge what is true of its design and framework or not, it still "speaks" on its own accord of its truth.[1]

These two images are a representation of Earth's axial shift from true north (before destruction of the vaulted dome by the explosion) to Earth's current axial rotation without the vaulted dome after the explosion.

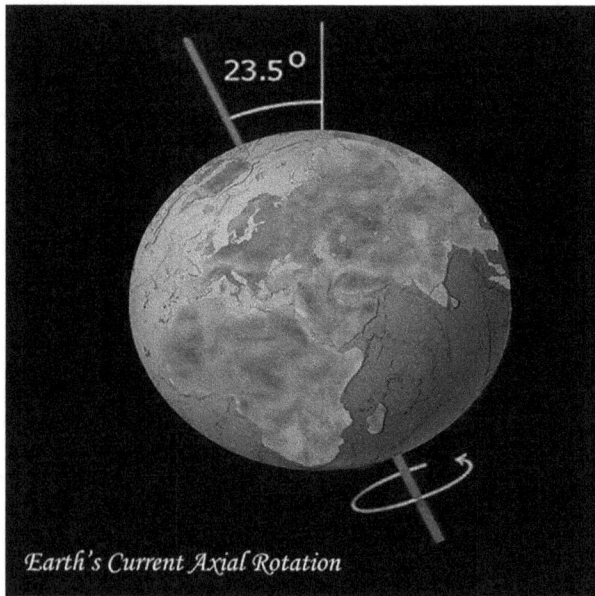

Earths Current Axial Rotation

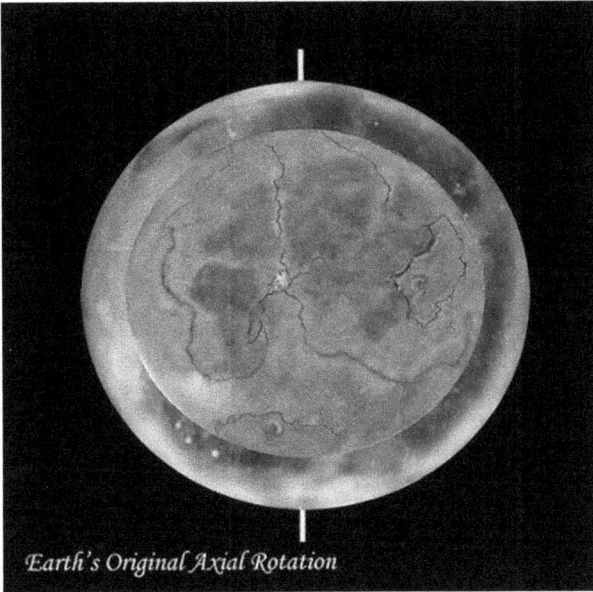

Earth's Original Axial Rotation

Orig Axial Rotation

A Spreading Mid-Ocean Ridge or a Cleft Channel?

The "so-called" spreading mid-ocean ridge encircling our globe is a sediment-filled canyon. This globe-encircling channel, carved within the face of the deep at the ocean floor, is accurately described by Proverbs 8:27: "When He inscribed a circle on the face of the deep." The word "inscribe" means to carve or engrave within a surface, another verse substantiating this concept of inscribing or carving something is *who has cleft a channel for the flood"* (Job 38:25). The word cleft means to part or split as if by cutting; the word "cleft" represents a hollowed-out area of indentation.

107

Like the seam of a baseball, the mid-ocean rift valley wraps around the globe and is crossed by intersecting right angles. Holy Scripture describes these right angles as cornerstones: "Where were you when I laid the foundation of the Earth? [. . .] On what were its bases sunk? Or who laid its corner-stone" (Job 38:4-6). These verses of Holy Scripture present a very accurate picture of the foundation of Earth at the ocean floor. The possibility of the ocean floor bearing witness to an excavation of the global supercontinent by fountains exploding from the great deep is worthy of consideration. Such an explosive process would expose the remaining substructure of chambers remaining at ground zero from where the channels of water burst fort

The strategic isolation, extreme terrain, and symmetrical architecture of the Mid-Ocean Ridge suggest it is the remaining substructure of "the storehouses of the deep" (Psalm 33:6-7).

The Mid-Ocean Rift does not necessarily represent a diverging boundary, a spreading center, or convection zone as many geologists and scientists believe. The claim of a minutely detectable measure of spreading between continents is not based upon "observable" evidence. It should not be denied the Earth is in a continual state of dynamic change. However, the simulation of sea-floor spreading does not result in the present geologic features of Planet Earth. The idea of an isolated supercontinent called "Pangaea" being spread apart by an upwelling asthenosphere—as magma pushed up from volcanic fissures and convection zones within the supercontinent—this geologic activity

108

would result in the formation of an extrusive volcanic ridge above sea level, within the supercontinent. To suppose such an extrusive, igneous structure would somehow cave in upon itself and end up at the sea floor millions of years later as a gaping canyon is mere speculation. The towering vertical pillars we see at the ocean floor do not lend any credence whatsoever to this concept.

In contrast, Genesis 7:11 suggests the supercontinent was transformed by an enormous blast when all the fountains of the great deep *burst* forth. If we take into consideration that (a) ocean basins were excavated by a great explosion from chambers hidden beneath the formless surface of Earth, and (b) the remaining landmass adjacent to the ocean basins is contained by an unbroken boundary known as the continental margin; with these considerations it becomes evident diverging or converging of ocean floors and/or continents did not take place. From a global perspective, the geology of Earth provides evidence it was transformed by dynamic vertical relief and not the speculated horizontal movement of continents.

An explosion of super-compressed steam would have excavated the ocean basins, leaving the pillar-like foundation of Earth exposed at the ocean floor. These are intrusive, basalt pillars forming the global foundation of Earth. "The mid-ocean ridges extend throughout the world's oceans and appear in places to continue beneath continents."[1] According

---

[1] M.J. Keen, *An Introduction to Marine Geology*, (New York: Pergamon Press, 1968), 138.

to Holy Scripture, "the pillars of the Earth are the Lord's, and He set the world on them" (Samuel 2:8). According to geologist Lester King, it is unmistakable that vertical relief was the instrumental geodynamic process responsible for the geologic formations we observe at the surface of our planet today.[1]

Have you ever thought to question how it was possible for rain to continue falling over forty days and forty nights until it once again covered Earth's formless surface? "And the rain fell upon the Earth for forty days and forty nights" (Genesis 7:12). Remaining portions of the vaulted dome began to melt as steam from the explosion continued having an impact upon this great sphere of ice surrounding our planet. As the vaulted dome melted, it began to rain for forty days and forty nights. The Bible tells us it was immediately following the explosion and this enormous rainfall began as "the floodgates of the sky were opened" (Genesis 7:11b). After forty days, the Bible tells us that approximately twenty-three feet (15 cubits) of water covered the entire planet for 110 days. Then the water began to recede for five months (150 days). "And the water receded steadily from the Earth, and at the end of one hundred and fifty days the water decreased" (Genesis 8:3). We may want to question, where did all of this water go?

Water as weight-in-motion flowed into the newly excavated ocean basins, exerting pressure upon the adjacent landmass and the cooling foundation of Earth. As the heated mantle and foundation of Earth settled and began to crystallize,

---

[1]Lester C. King, *Wandering Continents and Spreading Sea Floors on an Expanding Earth*, (New York: John Wiley & Sons Ltd., 1983), 24.

they were impacted by the tremendous weight of sediment-laden water flowing into recesses and cavities formed by the explosion. The depth and weight of water collecting in these newly excavated regions began to create a momentum of hydraulic pressure. It was this process of weight in motion that caused enough tension to plunge the Mariana Trench down to depths of 36,000 feet.

Portions of the Earth's mantle would continue outgassing as volcanic plumes, resulting in a voggy atmosphere. Unstable sections of the heated foundation were malleable and responsive to the dynamics of reciprocal equilibrium. Isostatic adjustments and vertical relief transformed our once formless planet into what we observe of Earth's features today, towering mountains and oceanic trenches sinking to unimaginable depths. After the flood, a large portion of water collected within abyssal trenches skirting the perimeter of the Pacific Ocean basin.

As these dynamic forces of reciprocal equilibrium were at work, low-pressure areas at the base of intrusive mountain ranges rebounded with isostasy. Psalm 104:6-8 accurately records that when the flood waters receded, this was when the mountains began rising to penetrate the overlying layers of strata. It was this dynamic of geological processes that created compressional and tensional stress, causing the deformation and displacement of deposited strata. As the mountains rose, great depths of the Earth were thrust up toward the surface, creating faults and inconsistencies within the layers of strata. "The mountains rose; the valleys sank down to the place which Thou didst establish for them" (Psalm 104:8).

Ark Above Ararat

CHAPTER 8

THE MOUNTAIN TOPS APPEARED

Geologist Lester King was correct with his assertion that the predominant features of Earth reveal vertical tectonic relief. Genesis 8:4-5 tells us the ark landed "above" the mountains of Ararat in the seventh month, and it wasn't until three months later in the tenth month that the tops of the mountains became visible: "And the water decreased steadily until the tenth month; in the tenth month, on the first day of the month, the tops of the mountains became visible" (Genesis 8:5). Misunderstanding this verse led people to visualize the tops of mountains becoming visible out of the water. But Scripture clearly states the water had been receding for three months. If the mountains became visible out of the water, the ark would have been floating and not resting. Genesis 8:5 is revealing a very important fact. The mountain tops became visible as they were pushed upward out of the ground by the forces of reciprocal equilibrium. Three months after the ark landed upon saturated ground above the mountains of Ararat, the tops of the mountains became visible. The fact that the tops of the mountains became visible in the tenth month signifies the intrusive mountain batholiths rose out of the ground.

> *Thou didst cover it with the deep as with a garment; the waters were standing above the mountains. At Thy rebuke they fled; at the sound of Thy thunder they hurried away. The mountains rose; the valleys sank down to the place which Thou didst establish for them* (Psalm 104:5-8).

King David's Psalm is a direct reference to the epic year of the flood because he clearly states the water would not return to cover the Earth. This Psalm could not have been written in reference to creation day three, because water did return to cover the entire Earth after creation week. The mountains rose and the valleys sank at the time of Noah's flood as water covering Earth's formless surface collected within the ocean basins and deeply excavated trenches. It was a time when extreme hydraulic pressure was exerted upon eroding banks, and as the saturated banks caved in, abyssal plains were formed. Volcanic eruptions continued; submarine landslides and the erosional effects of water continued widening the breach of excavated ocean basins.

As water receded off the remaining landmasses to collect at the sea floor and within abyssal trenches, hydraulic pressure increased and thrust tension upon the roots beneath the cooling foundations of great batholiths of granite. This balancing act facilitated the rise of intrusive, igneous mountain ranges in correspondence with sinking trenches. The hydraulic high-pressure areas of abyssal trenches exerting reciprocal equilibrium upon low-pressure areas is what elevated mountain ranges and ridges. This dynamic measure of weight in the balance was the origin of mountain ranges.

Who has measured the waters in the hollow of His hand, and marked off the heavens by the span, and calculated the dust of the Earth by the measure, and weighed the mountains in a balance and hills in a pair of scales? (Isaiah 40:12)

After the ark landed, Earth was still trembling. The great ship lifted up as the mountains of Ararat rose from beneath the ground. This is why it would take another seven months before Noah and his family were commanded by God to leave the safety of the ark. They remained safe within their ship of salvation until the unstable segments of Earth's crust finished rebounding with vertical relief.

The Earth-transforming event of the great flood began on the seventeenth day of the second month. It was 377 days after they had entered the ship that God gave Noah and his family permission to leave the safety of his ark: "And in the second month, on the twenty-seventh day of the month the Earth was dry" (Genesis 8:14). Noah let down the ramp and every species, along with the eight people who survived this catastrophic event, were finally allowed to disembark. As they left behind the ark of their refuge, a whole new world was laid before them as they viewed Mount Ararat towering 16,000 feet above sea level. Their dwelling here on Earth was no longer a smooth, formless surface, and the first rainbow emerged because the vaulted dome was no more.

Excavated Ocean Basins and the Continental Boundary

The steam explosion from the great deep subterranean chambers excavated ocean basins and left in its wake all land above sea level contained within "the unbroken boundary of a continental margin." Today the continental margin is a

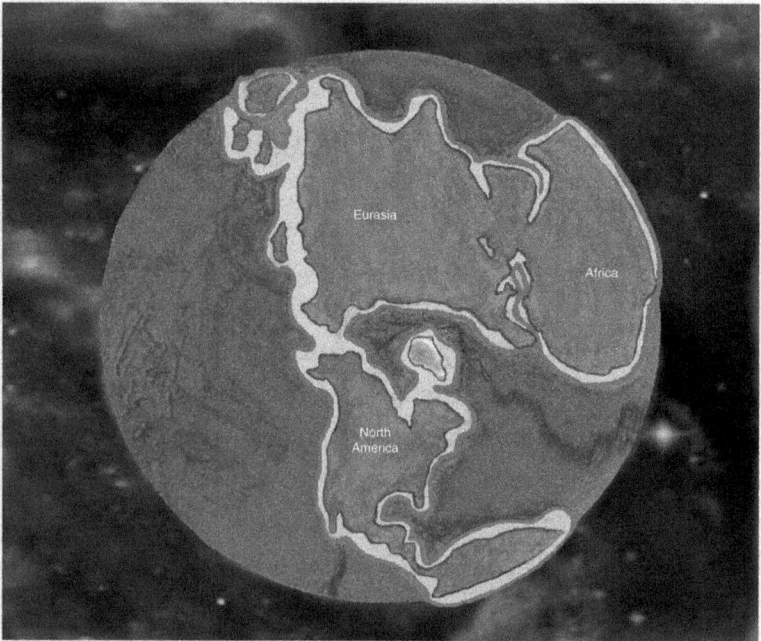

Continental Margin

submerged boundary about three hundred feet below sea level. It separates the continental landmass from the deep-sea floor. "The true geologic margin of a continent—that is, where continental crust changes to oceanic crust—is below sea level"[1] It is an important geologic feature which discloses overwhelming evidence against continental drift because all land above sea level is outlined and contained within this continuous, unbroken boundary.

It is the girth uniting seven major provinces as one super-continent above sea level and presents us with significantly profound evidence for the concept that Earth is firmly established upon a solid foundation. The existence of this boundary opposes any idea of continents sliding across the face of Earth horizontally. The importance of this boundary would have been overlooked without the guidance of Holy Scripture's reference to this amazing feature in Psalm 104:9: "Thou didst set a boundary that they may not pass over; That they may not return to cover the Earth."

I invite you to follow the continuous boundary of the continental margin using Google Earth or a satellite map of the world. It will be a submarine tour because the continental margin exists just below sea level. We will begin at the Bering Sea where Alaska is joined to the Kamchatka Peninsula and the continental margin drapes across the northern extremity of the Pacific Ocean uniting Asia with North America. From here we travel eastward along the coast of the United States toward South America where the margin narrows. Adjacent to the narrow margin of South America is the Atacama Trench, also known as the Peru-Chile trench system, which

---

[1]James S. Monroe and Reed Wicander, *Physical Geology – Exploring The Earth*, (New York: West Publishing Co., 1992), 319.

dives 26,000 feet below sea level for a stretch of 3,000-plus miles.

The continental margin winds around the tip of South America and stretches east over the Falkland Plateau. From here it continues to envelope the continent of Antarctica after winding through the islands of the Scotia Ridge, South Sandwich Islands, and South Orkney Islands. Along the east coast of South America the broad continental shelf proceeds northward, encompassing Central and North America, then the east coast of Greenland, where it enters the Arctic Basin and loops around to continue south along the west coast of Norway, Great Britain, and Spain.

Continuing past the Rock of Gibraltar to Africa, the continental margin wraps around the tip of South Africa and proceeds north toward the Gulf of Aden. The Red Sea and Gulf of Aden are both contained by the continental margin before it sweeps beneath the Arabian Peninsula and into the Gulf of Oman. The unbroken boundary continues along the coast of Iran, Pakistan, India, and Sri Lanka.

It is of significant interest that the island of Sri Lanka off the southern tip of India is contained within the continental margin. The margin continues along the Java Ridge Archipelago where it converges with the continental shelf north of Australia. The broad continental shelf of Indonesia and its many islands unite the mainland of Asia to Australia, just as Sri Lanka is united by the continental shelf to India. The Philippine Archipelago lies at the eastern extension of Indonesia's continental shelf and sea. The continental margin coincides with the Philippine Trench, which merges with the Ryukyu Trench and Japan Trench, skirting the coastline of the Japanese Archipelago. The Japan Trench merges with the Kuril Trench and shoots north to merge with the Aleutian Trench off the coast of Kamchatka.

This completes our circumterrestrial tour of the uniform, unbroken, continental margin. It is a steeply eroded continental rise and trench system defining the perimeter of excavated ocean basins. The continental rise is a low-relief zone of accumulated sediments that lies between the continental slope and the abyssal plain. It is a major part of the continental margin, covering around 10 percent of the ocean floor.[1] The existence of a continuous continental margin uniting all land above sea level is perhaps the most compelling geologic evidence against horizontal movement of continents over the surface of Earth and confirms the Scriptural concept of a firmly established Earth whose crust was ruptured by all the fountains of the great deep bursting forth.

> **"When He set a boundary for the sea,**
> **so that the waters would not surpass His command,**
> **when He marked out the foundations of the Earth."**
> —Proverbs 8:29 NBS

---

[1] "Continental rise," https://en.m.wikipedia.org/wiki/Continental_rise.

Fossils and Fossil Fuels

As I was leafing through a pile of books at the University of Oregon Science library, I discovered tektites, which I had never heard of before. My curiosity was spurred as I read theories about how they had been formed by giant meteors crashing into the Earth. But there seemed to be a problem with this hypothesis because researchers were unable to locate impact craters in the surrounding vicinity for many kinds of tektites scattered all over the globe. And more specifically, they were unable to locate meteor impact zones for the creation of the Australasian and Libyan Desert Glass kinds of tektites.

Tektites are predominantly made of silica, but they vary to include minerals from their specific region of source. I propose an unmitigated, catastrophic, silica-rich explosion as is described by Genesis 7:11 most certainly would have had the potential to produce conditions for the formation of a variety of tektites. The nature of this dynamic event would have scattered them all over the planet. The flanged button tektites of Australia were formed aerodynamically as intense heat caused them to become molten glass while shooting through the atmosphere. These dynamics formed this rock into a symmetrically shaped rimmed button.

The destructive power of a global hydrothermal catastrophe, like the steam explosion that took place when all of the fountains of the great deep burst forth, an event like this would have geologically resulted in Earth's crust being composed of the simplest forms of silicate minerals being chemically bonded with more complex minerals because heat, pressure, and water were the dynamic forces required for this type of transformation.

Earth's surface is composed of a multitude of complex minerals, and yet what we observe happening with present-day processes is corrosion and decomposition. We are not seeing complex minerals being created. In fact, what we observe today is complex minerals degrading into less complex forms. This observation leads to another question; what dynamic conditions were in place to create such complexity? The physical and chemical laws of nature have remained the same and maintain a revelation of truth.

The physical characteristics and chemical composition of complex metamorphic minerals is determined by the varying degrees of contact they have had with fluid, heat, and pressure. About one-third of all known minerals include the element silica, and "silicate minerals [...] make up perhaps as much as 95% of Earth's crust."[1] Geologist Lester King points out that siliceous, superheated steam impacted a variety of mineral forms, "by comparison with adjacent rocks the levity appears to be due to associated volatiles of which siliceous, superheated steam is usually the most prominent member."[2] It seems likely that blasts of siliceous superheated steam

---

[1] James S. Monroe and Reed Wicander, *Physical Geology – Exploring The Earth*, (New York: West Publishing Co., 1992), 65.

[2] Lester C. King, *Wandering Continents and Spreading Sea Floors on an Expanding Earth*, (New York: John Wiley & Sons Ltd., 1983), 7.

would also result in petrified forests. Perhaps the formation of coal took place in regions where trees were exposed to blasts of penetrating, superheated, siliceous steam before being trapped deep beneath the exploded caprock. Logs sealed under deposited layers of rubble would have transformed into coal when exposed to residual heat radiating from Earth's disgruntled foundation. These types of phenomena are so radical that most scientists refuse to acknowledge it was a single event, a catastrophe that was recorded by all tribes and nations throughout human history. Many educated people have been led to believe it was multiple events taking place over millions of years and refuse to even consider that Holy Scripture has anything to say about it.

Water as a catalyst for Noah's flood would end up with residue from the minerals and elements dissolved by super compressed steam and pressure. "Most of the Earth's crust is covered with ocean water that contains many elements in solution and in addition massive deposits of valuable minerals such as copper, manganese, nickel and cobalt have been discovered within the ocean ridges along the sea floors and within sea-floor sediments."[1] Present-day ocean water seems to be a fossil testament of Noah's flood. How did our oceans end up containing every mineral and element known to man within suspended animation? This may have been one of the questions leading scientist Gerry Bearman to think, "It soon became clear that hydrothermal activity must have been a

---

[1]James S. Monroe and Reed Wicander, *Physical Geology – Exploring The Earth*, (New York: West Publishing Co., 1992), 335.

major and previously unconsidered contributor to the chemical mass balance of oceans throughout the history of Earth."[1] The kind of immense hydrothermal activity required to produce volumes of oceanic-type water is nonexistent at present.

Layers of metamorphic rock, chemical sediments, complex minerals, sand, silt, clay, mud, and electrically charged particles of ions all are relevant to the Scriptural model of fountains bursting from the great deep. Remarkably, this is exactly what the collective sum of Earth's crust, ocean water, and atmosphere reveals. A global hydrothermal explosive release of pressure at the onset of the flood would result in chemical bonding, petrification, entrapment, and layered burial of life forms preserved as fossils, along with the creation of fossil fuels. Massive deposition and rapid entrapment of living organic things like shells, fish, animals, plants, and bodies, this was carbon material exposed to heat from below and pressure from above, the quintessential requirements for the formation of fossil fuels like coal, oil, shale, and natural gas.

Billions of gallons of hydrocarbons within oil and natural gas reserves are recovered annually throughout the world as a super-abundant energy resource. "Both petroleum and natural gas are hydrocarbons, meaning that they are composed of hydrogen and carbon."[2] Compounds that contain carbon are derived from organic source materials of plant

---

[1] Gerry Bearman, ed., *The Ocean Basins: Their Structure and Evolution*, (New York: The Open University, Pergamon Press, Inc., 1989), 96.

[2] James S. Monroe and Reed Wicander, *Physical Geology – Exploring The Earth*, (New York: West Publishing Co., 1992), 181.

or animal origin. "Most geologists believe that natural gas and oil originate from the decay of buried organic material, hence the name fossil fuels."[1]

According to Monroe[2] the conditions required for hydro-carbon/fossil fuel formation are:

1) A place where little oxygen is available to decompose the organic compounds.
2) Burial at depths where exposure to heat transforms them into petroleum and natural gas.
3) Entrapment by impermeable caprock preventing upward migration of the hydrocarbons.

In accordance with and to substantiate the plate tectonic hypothesis, scientists postulate the organic source materials for hydrocarbon formation came from microorganisms such as plankton and kelp sinking to the ocean floor, where there is little oxygen. Here this organic material is supposedly covered by layers of sediment. "Then as sediments accumulate over organic layers, pressure, temperature, and time break these organic molecules into the heavy hydrocarbons of oil."[3] Question is, how did organisms so tiny migrate from

---

[1]Richard Monastersky, "Drilling into a deep controversy," *Science News* 131, no.24 (June 1987): 380-381.

[2]James S. Monroe and Reed Wicander, *Physical Geology – Exploring The Earth*, (New York: West Publishing Co., 1992), 182-183.

[3]Richard Monastersky, "Drilling into a deep controversy," *Science News* 131, no.24 (June 1987): 87.

the seafloor and get deposited as vast fields of oil without any other contaminants? The accumulation, burial, and entrapment of massive deposits of biological life forms required to produce organic hydro-carbon fossil fuels is currently not observed to be taking place on a global scale. What we do see with present-day processes is dead animals and plants are subject to rapid decomposition and decay. They convert back to dust or are eaten by scavengers. Only massive deposition and rapid entrapment of organic life would result in huge deposits of uncontaminated fossil fuels.

The catastrophic global transformation documented by historic records as the Noahic Deluge outlines a fascinating blueprint for understanding the geo-physical and chemical processes involved in the formation of fossils. Creatures instantly trapped by a heavy mass of heated material were destined to leave behind fossil imprints within the geologic column. High-velocity winds and currents from the greatest explosion in human history resulted in the rapid formation of strata layers. Successive cooling, shifting, settling, and the vertical uplift of mountain ranges resulted in the unconformity of layers. The accumulation, burial, and entrapment of biological life forms resulted in great deposits of carbon, the source material of fossil fuels.

Fossil fuels were formed when pockets of plant and animal matter were instantly trapped beneath tons of rocks, dirt, and gravel and subsequently exposed to intense heat from Earth's flaring core (Genesis 7:11). Heat from below and pressure from above is what resulted in the creation of billions of gallons of uncontaminated crude oil. "Most geologists believe that natural gas and oil originate from the decay of buried organic material, hence the name fossil fuels."[1] It is extremely unlikely that oil fields were slowly formed over millions of years with the migration of minute particles of organic matter without any trace of contamination.

Perhaps the vast oil reserves of Saudi Arabia offer the most objective evidence for a cataclysmic event like Noah's flood, for this location was the hub of civilization prior to the event. Instant burial and entrapment of life at this once-thriving location is what provided a great volume of organic material for the creation of enormous oil and natural gas reserves.

Fossils are the result of living organisms being suddenly trapped beneath layers of rock, dirt, and gravel, an event we do not observe to be happening today. Dead animals do not sink (they float), and fossils are not formed by gradual deposits of soil due to erosion. The phenomena of fossil creation is the result of a sudden historic event. Remember the Lord did not say He would drown all living creatures but that He would "blot" them out by covering them with the Earth! "Thus, he blotted out every living thing that was upon the face of the land, from man to animals to creeping things and to birds of the sky, and they were blotted out, from the Earth; and only Noah was left, together with those that were with him in the ark" (Genesis 7:23). Another very important observation to notice is if each deposited layer of strata was to signify millions of years—as evolutionists suggest—we

---

[1] Richard Monastersky, "Drilling into a deep controversy," *Science News* 131, no.24 (June 1987): 380-381.

could expect to see an immense buildup of topsoil, but what we observe is an erosion so immense that "half of the topsoil on the planet has been lost in the last 150 years."[1]

The gradual transformation of Earth over billions of years may not be taking place at an invisibly slow pace. There is no evidence that a slow, gradual mechanism played any kind of role in the formation of most geologic features. Geologists use the continent of India as another example to substantiate the plate tectonics theory, continental drift, and mountain building. They suggest the continent of India broke away from Antarctica, floated across the spreading center/divergent boundary of the Mid-Oceanic Ridge and migrated north along the perfectly straight Ninetyeast Ridge before colliding with Asia. The continental collision proposed by this theory is what they believe forced the Himalayas to rise 29,000 feet above sea level.

The mobility of Earth's crust is hyped and promoted as fact, but Holy Scripture consistently tells us the Earth is stable. "The world is firmly established; it cannot be moved" (Psalm 96:10), and "He established the Earth upon its foundations, So that it will not totter forever and ever" (Psalm 104:5). Several geologists agree it would be impossible for many geologic features to have been created by shifting plates. "Indeed, the modern paradigm of plate tectonics would have the Arabian Peninsula drifting away from the African continent. The early formation of the Red Sea

---

[1] *World Wildlife Fund,* https://www.worldwildlife.org/threats/soil-erosion-and-degredation.

is somewhat obscure and is not clearly a result of sea-floor spreading."[1] The concept of mountain ranges being uplifted by a horizontal collision is also challenged by geologists who have seen perfectly aligned riverbeds in the Himalayas. "They have not suffered the slightest tilt. This line of evidence would indeed, be impossible to reconcile with the horizontal compression envisaged by the collision concept."[2]

Another theory attempting to confirm continental drift asserts different sections of the complex mountainous region of India were uplifted, folded, and plunged together over millions of years. However, every fold belt studied in India was discovered to have been formed by compressive stress in a north-south direction. The significance of this discovery is it is extremely unlikely that a north-south direction of stress would have remained constant for millions of years. Dubey (1986) and Bhat (1983) believe the size of their study demands a better explanation. "The time spans involved are large enough to demand an explanation in the total perspective of Earth's dynamics."[3] Without any manipulation of geologic data, vertical uplift of mountain ranges like the Himalayas can be understood from a biblical perspective.

---

[1] R. G. Coleman, "Geologic Background of the Red Sea," *The Geologry of Contnental Margins*: 743-751. https://link.springer.com/chapter/10.1007/978-3-662-01141-6_55.

[2] A. Bartos-Kyiakidis, *Critical Aspects of The Plate Tectonic Theory Vol. 1*, (Athens: Theophrastus Publishing & Proprietary Co., S.A., 1990), 134.

[3] A. Bartos-Kyiakidis, *Critical Aspects of The Plate Tectonic Theory Vol. 1*, (Athens: Theophrastus Publishing & Proprietary Co., S.A., 1990), 151.

Our world's oceans are approximately 2.5 times the size of combined land areas, and deep-ocean sedimentary plains make up about half the area of the entire globe. "In between ridges and trenches lie sediment covered abyssal plains."[1] Massive erosion and slumping of sediments formed the sedimentary alluvial fans adjacent to India. The floor of the Arabian Sea is also covered by an extensive sedimentary fan known as the Indus Cone, and the Bengal Fan is the largest submarine fan on Earth, covering 1,900 by 890 miles within the Bay of Bengal. These huge deposits have been analyzed and studied by geologists, who discovered layers of silt, gravel, clay, micas, crystalline carbonate, and opaque grains in the alluvial fans. Curiously they also contain plant fauna and lines of lamination, indicating high-velocity winds and currents were instrumental for their rapid formation. Once again, evidence reveals there was absolutely nothing slow about the geologic processes creating these stunning formations. These alluvial fans had to have been formed by a sudden, catastrophic event.

Layers of geologic strata reveal a cataclysmic process of rapid deposition took place at the Earth's surface as well. Smooth lines of lamination between layers of deposits extending for thousands of square miles clearly indicate water and high-velocity winds played a role in the formation of stratum.

Momentous exposure of the pillar-like foundation of Earth at the ocean floor—layer upon layer of sedimentary deposits

---

[1] Gerry Bearman, ed., *The Ocean Basins: Their Structure and Evolution*, (New York: The Open University, Pergamon Press, Inc., 1989), 28.

containing fossils—the upheaval of mountains—the sinking of valleys and trenches—all reveal themselves in the dramatic landscapes we observe above and below sea level today. These geologic formations reveal a moment in time like no other throughout all of history. The great flood of Noah lasted 375 days. For some time after this event had taken place, the Earth's climate was temperate and ocean water was very warm due to the explosive release of steam from the great deep underground chambers.

CHAPTER 9

NEW BEGINNINGS

"Now it came about in the six hundred and first year, in the first month, on the first of the month, the water was dried up from the Earth. Then Noah removed the covering of the ark, and looked and behold, the surface of the ground was dried up" (Genesis 8:13). God set a bow in the cloud as a sign of His promise and made a covenant with Noah along with every living creature for all successive generations to affirm that never again would He destroy the Earth with a flood of water (Genesis 8:21). This was the first time Noah had ever seen a rainbow expanding across the moisture-laden sunlit sky. The bow is mentioned *seven times* in Genesis verses 9:9-17. The rainbow is not only a reminder of God's promise to never again destroy the Earth with a flood, but it also signifies God's desire to have fellowship with the survivors and their ancestors.

Perhaps the reason for mentioning the rainbow multiple times is to draw our attention not only to God's promise but for us to realize how significant the change was for the Earth's climate. There was no rainfall prior to the flood, "but a mist used to rise from the Earth and water the whole surface

Leaving Noah's Ark

of the ground" (Genesis 2:6). The environmental requirements for the formation of a rainbow did not exist prior to
this point in time because of the vaulted dome. Loss of this
protective canopy not only resulted in rainfall but a radical
shortening of lifespans for people and animals. Genesis 9:29
tells us Noah lived for 349 years after the flood and died at
950 years of age. "The first three generations born after the
flood lived 400 plus years. The fourth generation and the
generations afterwards [Peleg and his descendants] lived
only a little more than 200 years."[1]

---

[1]Ken Johnson, *Ancient Post-Flood History*, (2010), 44.

After the vaulted dome was destroyed, people lived much shorter lives, but the population boomed because of a global temperate climate prevailing, and seeds sprouted from the abundance of fertile soil deposited by this catastrophic event. Holy Scripture tells us Noah began farming and planted a vineyard (Genesis 9:20). In 2011 the *Washington Post* published an article by Marc Kaufman about a 6,000-year-old winemaking operation within the Areni caves, just sixty miles from Mt. Ararat. The *Post* quotes archaeologist Gregory Areshian as saying, "This is the oldest confirmed example of winemaking by a thousand years [. . .] People were making wine here well before there were pharaohs in Egypt."[1] Not only did they discover an advanced winemaking operation, but a laced leather moccasin was in the same cave. This shoe is displayed at the History Museum of Armenia and may have belonged to Noah, according to the timeline measured by the chronology recorded in Holy Scripture.

The oldest leather footwear in the world is believed to be 5,500 years old, and it was discovered at The Cave of Areni in 2008. "It is a one-piece leather-hide shoe, the oldest piece of leather footwear in the world known to contemporary researchers." In the same cave they discovered "the earliest known wine press for stomping grapes, fermentation and storage vessels, drinking cups, and withered grape vines,

---

[1] Marc Kaufman, "Ancient winemaking operation unearthed in Amernian cave," *The Washington Post*, (January 2011), https://www.washington post.com/national/ancient_winemaking_operation_unearthed_in_arme nian_cave/2011/01/06/ABhQHkD_story.html.

skins and seeds."[1] The same article goes on to say, "Patrick McGovern, a biomolecular anthropologist at the University of Pennsylvania, commenting on the importance of the find, said, 'The fact that winemaking was already so well developed in 4000 BC suggests that the technology probably goes back much earlier'"[2]

The Bible records that one day, Noah drank so much of his delicious wine it caused him to fall asleep naked in his tent. When his youngest son, Ham, saw his father without clothes on, he ridiculed him as he shared with his brothers what he had seen. Noah woke up knowing his son had been disrespectful and cursed him along with all of his descendants, saying, "Cursed be Canaan, a slave of slaves shall he be unto his brothers" (Genesis 9:25). Noah's curse may be the reason why his "great" grandson Nimrod was rebellious and contemptuous. Legend has it that Ham gave his grandson Nimrod the sacred garments of Adam, which endowed him with great physical strength and wisdom, and as a result, Nimrod grew to be a mighty hunter with rulership over several kingdoms in Shinar.

In the ancient world, immediately after the great flood of Noah, all of humanity had common roots and were united as a family. Our roots run deep, and we truly are brothers and sisters of a noble human family, but the devil is in the details

---

[1]Shereen Dindar, "World's Oldest Leather Shoe Found—Stunningly Preserved," *National Post Canada*, (June 2010), https://www.historyof information.com/detail.php?entryid=3197.

[2]"Areni-1 winery," https://en.m.wikipedia.org/wiki/Areni-1_winery.

with our freedom to choose. Originally it was not about the devil being in the details, as if this was a burdensome, judgmental thing. It was about God being in the details, implying that whatever you do should be done accurately, truthfully, and thoroughly because taking care of the details is extremely important.

The records of Genesis chapter ten names the descendants of Noah, his three sons: Japeth, Ham, and Shem. These names unlock the mystery of history with biblical stories about places they settled. For example, Japeth had a son named Gomer whom he named Togarmah (Genesis 10:3).

> Togarmah had eight sons. Their names are recorded in History of Armenia 1:1; and the Georgian Chronicle 1:1. The firstborn son of Togarmah, called Hayk by the Armenian people, lived in the area today known as Babylon ... The History of Armenia continues explaining that after the birth of Hayk's firstborn son, Armaneak [whom the Armenian people claim as their founder], Nimrod took control of the known world and formed the first post-flood empire. While in service to Nimrod, Hayk saw the government become more and more corrupt. Hayk decided to take his family and leave for a place far away from the land of the evil empire.[1]

Thus the nation of Armenia was named after Hayk's firstborn son, Armaneak, the grandson of Togarmah (a.k.a. Gomer), and great grandson of Japeth.

---

[1] Ken Johnson, *Ancient Post-Flood History*, (2010), 64-65.

Nimrod and the Tower of Babel

The land of Shinar is mentioned several times in the Bible. The Tower of Babel was built on "a plain in the land of Shinar" (Gen 11:1-9), and Genesis 14:1 and 9 tell us Shinar was the home of King Amraphel. "Genesis Rabbah 42 says Amraphel was called by three names: Cush, after his father's name (Gen. 10:8), Nimrod, because he established rebellion (mrd) in the world, and Amraphel, as he declared (amar) 'I will cast down' (apilah)."[1] The Bible tells us Amraphel defeated many giants and conquered their territory along with four other kings, one of them being Chedorlaomer (Genesis 14:5-7). Giants roamed the Earth once again after the flood of Noah. I am curious to discover how their bloodline was carried forward. Perhaps it was by one of the wives of Noah's three sons.

Nimrod is credited with building one of the first major cities after the flood. There was only one language spoken as generations from Noah and his three sons multiplied. God told them to scatter and populate the Earth (Genesis 9:1). Some of Noah's descendants chose to obey God's commandment and dispersed, but the rest chose to follow the strong and mighty leadership of Nimrod. They followed him east to the land of Shinar, which some scholars believe is more or less equivalent to the Mesopotamian "Sumer and Akkad."[2] The Jewish Encyclopedia states, "Nimrod is the prototype of a rebellious

---

[1] "Amraphel," *Wikipedia*, https://en.wikipedia.org/wiki/Amraphel.

[2] Yigal Levin, "Nimrod, Mighty Hunter and King — Who Was He?" *The Torah.com* (2020). https://www.thetorah.com/article/nimrod -mighty-hunter-and-king-who-was-he.

people, his name being interpreted as [. . .] he who made all the people rebellious against God."[1] "Nimrod took control of the known world and formed the first post-flood empire."[2]

There are four Genesis verses referring to Shinar as the place where the original Tower of Babel was built—Genesis 10:10, 11:2, 14:1, and 14:9. It is estimated that Nimrod ruled until 2218 BC, and his construction of the Tower of Babel was a grandiose insult. Not only did people refuse to obey God's edict to populate the whole Earth, but they chose to exalt Nimrod as their god and made the same mistake as Adam and Eve, believing they should be glorified like God (Genesis 3:5). Josephus wrote:

> *Now it was Nimrod who excited them to such an affront and contempt of God. He was the grandson of Ham, a bold man, and of great strength of hand. He persuaded them not to ascribe it to God, as if it were through His means they were happy, but to believe that it was their own courage which procured that happiness. He also gradually changed the government into tyranny, seeing no other way of turning men from the fear of God, but to bring them into a constant dependence on his power. He also said he would be revenged on God, if he should have a mind to drown the world again; for that he would build a tower too high for the waters to reach. And that he would avenge himself on God for destroying their forefathers.*[3]

---

[1]Pes. 94b; comp. Targ. of pseudo-Jonathan and Targ. Yer. to Gen. x. 9
[2]Ken Johnson, *Ancient Post-Flood History*, (2010), 65.
[3]Josephus Flavius, *Antiquities of the Jews*, (Blacksburg: Unabridged Books, 2011), Loc. 59.

Now the multitude were very ready to follow the determination of Nimrod, and to esteem it a piece of cowardice to submit to God; and they built a tower, neither sparing any pains, nor being in any degree negligent about the work and, by reason of the multitude of hands employed in it, it grew very high, sooner than anyone could expect; but the thickness of it was so great, and it was so strongly built, that thereby its great height seemed, upon the view, to be less than it really was. It was built of burnt brick, cemented together with mortar; made of bitumen, that it might not be liable to admit water. When God saw that they acted so madly, he did not resolve to destroy them utterly, since they were not grown wiser by the destruction of the former sinners; but he caused a tumult among them, by producing in them diverse languages, so they should not be able to understand one another. The place wherein they built the tower is now called Babylon, because of the confusion of that language which they readily understood before; for the Hebrews mean by the word Babel, confusion. . . ."[1]

Nimrod's success in getting people to worship the stars as icons of immortality was a great deception. His rebellion is evidenced by his desecration of the true significance of the stars God placed in the heavens to tell of His glory. The influence of Nimrod contrasted sharply with that of Seth, who used the stars to tell God's story. Astronomer E.W. Maunder reveals how star constellations align with and tell the story of Genesis

---

[1]Josephus Flavius, *Antiquities of the Jews*, (Blacksburg: Unabridged Books, 2011), 1.115.

in his book *The Astronomy of the Bible*.[1] Maunder explains the bow and arrow of Sagittarius represents God's bow in the cloud and His promise to never again destroy all life on Earth with a flood of water (Genesis 9:15). He goes on to describe how the constellation Virgo represents the Virgin mother who would immaculately conceive the Son of God. How the constellation Leo represents the "Lion" of the Tribe of Judah (Revelation 5:5). "The heavens are telling of the glory of God; And their expanse is declaring the work of His hands" (Psalms 19:1). "Let there be lights in the expanse of the heavens [. . .] let them be for signs, and for seasons" (Genesis 1:14). Why look to the stars? I say look to the Creator of the stars who paid a great price for the redemption of your soul. He provided a way for the restoration of our soul, and just like Adam and Eve in the Garden of Eden, we are privileged to walk and talk with Him. The Creator of the stars will guide us in paths of righteousness!

The apostasy led by Nimrod, in which the creation was worshiped rather than the Creator, continues to dominate our world today. Not only did Nimrod get people to worship the stars as icons of immortality but he succeeded in having people worship himself as god almighty. His mother/wife (Semiramis) along with their child (Tammuz) were established by Nimrod to be honored and worshiped as a holy trinity. Nimrod schemed and propagated an evil doctrine promoting his wife as the "Queen of Heaven" and their child Tammuz, born on December 25th, as the "divine son of heaven." This

---

[1] E. Walter Maunder, F.R.A.S., *The Astronomy of the Bible*, (Sai ePublications, 2014), 30.

was the plan of Satan, brought into manifestation through Nimrod. This counterfeit system of worship has "unknowingly" prevailed until now. The overall cultural acceptance of Christmas is the holy cow of our time. It is a counterfeit system of worship wrapped within layers of deception intended not only to oppose the true God of heaven but to make a mockery of the sovereign nature of who God is.

Generation after generation worshiped Tammuz, who was a false Messiah, son of Baal, the "Sun God." Holy Scripture records that 1,400 years after the reign of Nimrod, his influence persisted: "Then he brought me to the entrance of the north gate of the house of the LORD, and I saw women sitting there, mourning the god Tammuz" (Ezekiel 8:14). The prophet Ezekiel was counseled by God that worship of the sun god Tammuz in the Jewish temple was an abomination. When Ezekiel entered the inner court of the Lord's house, what he saw at the entrance to the temple "between the porch and the alter, were about twenty-five men with their backs to the temple of the Lord and their faces toward the east, and they were prostrating themselves eastward toward the sun" (Ezekiel 8:16-17).

Author Amir Tsarfati vividly points out the same thing is happening today as a new form of religion is being promoted all across the world with music, pop-culture, tattoos, entertainment, and technology. People are flocking to great cities like Dubai to marvel over man's achievements. And now they plan for an even greater city in Saudi Arabia called Neom. "They are about to build a city that is one building that stretches for nearly 70 miles, to make everything green and nice, untouched, with a perfect climate. It will attract billions of dollars and people will be worshiping not just the rulers but even the city itself."[1] Neo-paganism is the new religion of today, which people believe stems from pre-Christian beliefs. I agree these beliefs are from a time before the Bible was written, but this doesn't mean they are pre-Christian, for God made His plan of redemption known from the very beginning through Adam.

Nimrod's grandiose building project was a display of extreme hubris intended to deceive his naive followers. At the culmination of their architectural achievement, God descended from heaven and saw their project was built to undermine His plan of salvation for human beings. With righteous anger, judgment was passed and people no longer understood each other as their one language became many. Construction ceased as the builders were unable to communicate (Genesis 11:7). "Therefore, its name was called Babel, because the Lord confused the language of the whole Earth; and from there the Lord scattered them abroad over the face of the whole Earth" (Genesis 11:9).

Some think the timeline for this event is revealed by Genesis 10:25, which tells us it was in the days of Peleg the Earth was divided. Peleg was born to Eber ninety-six years after the flood of Noah; he was a descendant of Noah's son Shem. This indicates that dispersion of people from the Tower of Babel took place about one-hundred years after the flood. The confusion of languages forced people to split up. The descendants of Noah migrated north, south, east, and west from the Plain of Shinar. As people settled, new regions in the course

---

[1] "Prophecy Roundtable - Mystery Babylon" *YouTube*, https://youtu .be/Z33finAjlKA.

of migration and adaptation to new surroundings, unique cultures developed. Jewish historians later used the genealogic information provided in Genesis chapter ten, along with the book of Chronicles, to create what is known as "The Table of Nations," a chart naming and tracking the ethnic diversity of seventy families who sprang from Noah and his three sons.

A Great Migration

Diversification was not only influenced by alteration of the once common language, but the extreme geologic and environmental variations of their new surroundings had a tremendous impact. The Earth was no longer formless, and it was a hardy breed of people exploring distant lands. Some trekked across high mountains and settled in what is now known as China. Stories were passed down from generation to generation about the long and arduous journey over the K'un-lun Mountain chain, with pinnacles towering more than 22,000 feet above sea level. This great mountain became known to future generations as the abode of the gods, where their ancestors drank from the fountain of immortality and wherefrom flowed four great rivers of the world.

The ancient Chinese believed in one God, a supreme heavenly being named Shang Ti whose name was written with the radicals for Heaven (above) and Emperor—thus "Heavenly Emperor." The *Chinese Book of History*, the Shu Jing, records that in the year 2230 B.C., the Emperor Shun "sacrificed to Shangdi" (Shang Ti). That is, he sacrificed to the supreme God of the ancient Chinese, Shangdi meaning "Supreme Ruler." The exact text of the ceremony is surprisingly similar to the first chapter of Genesis:

Of old in the beginning, there was the great chaos, without form and dark. The five elements [planets] had not begun to revolve, nor the sun and the moon to shine. In the midst thereof there existed neither forms for sound. Thou. O spiritual Sovereign, camest forth in Thy presidency, and first didst divide the grosser parts from the purer. Thou madest heaven; Thou madest Earth; Thou madest man. All things with their reproductive power got their being.

Over time people gradually lost an understanding of what this sacred ceremony was all about. "In China, a collective name for gods, perhaps representing one supreme god or overlord. Ti were worshiped as deified ancestors of the Shang dynasty, and the Shang rulers worshiped Shang-ti— but the absence of a plural form makes it uncertain whether Shang-ti was one or many."[1] The Shang Ti became obscured by all kinds of deities. At a much later date, Chinese beliefs branched out to Buddhism and Taoism. However, ancient records about migration of these indigenous peoples to the extreme eastern region of Asia is documented by the ancient Chinese character for the word "MIGRATE," containing the following four radicals: great + division + west + walking. "In

---

[1] s.v. "Shang-Ti," *Encyclopedia.com*, https://www.encyclopedia.com /religion/dictionaries-thesauruses-pictures-and-press-releases/shang-ti.

the west (Babel) there had been a great [. . .] division [. . .] This division resulted in their walking in a great migratory move."[1]

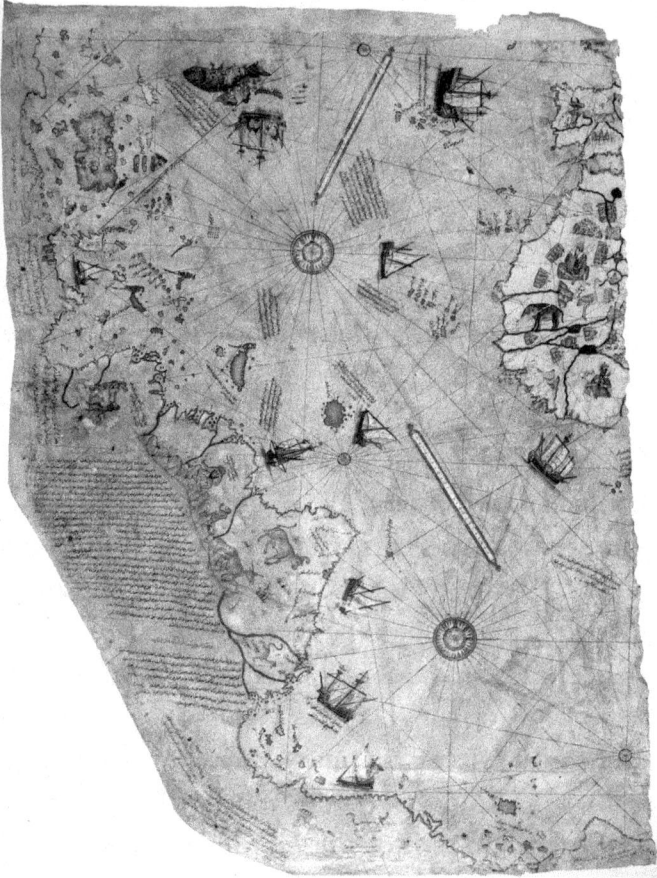

Surviving fragment of the Piri Reis map
Credit: https://commons.m.wikimedia.org/wiki/File:Piri__reis
_world_map_01.jpg#mw-jump-to-license

---

[1]C.H. Kang & Ethel R. Nelson, *The Discovery of Genesis: How The Truths of Genesis Were Found Hidden in the Chinese Language*, (Saint Louis: Concordia Publishing House, 1979), 109.

CHAPTER 10

OCEANIC NAVIGATION PRIOR TO THE ICE AGE

After the explosion of all the fountains of the great deep there remained a veil of smog blanketing the planet. Thick, volcanic clouds trapped moisture and heat from the explosion, resulting in an abundance of evaporation from the warm, temperate oceans. Sea levels were drastically lower than today due to evaporation, and there were land bridges between the continents. Animals and people migrated along these exposed coastlines to remote locations around the globe.

Noah's ancestors were expert seafarers with navigational knowledge handed down by Seth. They built ships, and Seth's wisdom of the stars provided a way for them to accurately map distant shorelines. Evidence for their expertise was discovered in 1929 as the Topkapi Palace of Istanbul was being transformed into a museum.[1] It was here they found an extremely rare map drawn on gazelle skin. Notations had been added to the map by Turkish admiral Piri Reis in the fifteenth century, but its true significance wasn't brought to light until a much later date.

It was in the 1960s when modern-day scientists began using Radio-Echo Sounding (RES) technology to map and view the coastline of Antarctica beneath ice two miles thick. This was when Captain Arlington H. Mallery—a student of old maps and borderland regions of archaeology, noticed a similarity between the coastline of Antarctica revealed by RES and the Piri Reis map. He made the following statement that in his opinion, "the southernmost part of the map represented bays and islands of the Antarctic coast of Queen Maud Land now concealed under the Antarctic ice cap. That would imply, he thought, that somebody had mapped this coast before the ice had appeared."[2]

Charles Hapgood pursued further investigation of the map's authenticity. After years of research with his students, along with support from the U.S. Air Division, Hapgood concluded the Piri Reis map must have been pieced together from prior maps charted with a knowledge of spherical

---

[1] https://commons.wikimedia.org/wiki/File:Piri_reis_world_map_01.jpg
Map of the world by Ottoman admiral Piri Reis, drawn in 1513. Only half of the original map survives and is held at the Topkapi Museum in Istanbul. The map synthesizes information from twenty maps, including one drawn by Christopher Columbus of the New World

[2] Charles H. Hapgood, *Maps of the Ancient Sea Kings: Evidence of Advanced Civilization in the Ice Age*, (Adventures Unlimited Press, 2014), 2.

trigonometry and magnetic declination. Hapgood asserts "the oblong grid, used by Ptolemy and found on the Piri Re'is Map, has its origin in an ancient use of spherical trigonometry." He also credits the ancient cartographers with "a scientific achievement far beyond the capacities of the navigators and mapmakers of the Renaissance, of any period of the Middle Ages, of the Arab geographers, or of the known geographers of ancient times. It appears to demonstrate the survival of a cartographic tradition that could hardly have come to us except through some such people as the Phoenicians or the Minoans, the great sea peoples who long preceded the Greeks but passed down to them their maritime lore."[1]

A continual coastline from South America to Antarctica can be seen on the Piri Reis map. Which is exactly what we would expect to see (a continuity of landmass) if the original map was drawn during the post-flood period when land bridges were exposed between continents due to low sea levels. Vast amounts of warm ocean water would have evaporated for years following the flood, lowering sea levels as much as 300 feet. After one hundred years of evaporation, the continental margin became exposed as a natural land bridge. What is now known as the Bering Sea was once known as the Land of Beringia.

Investigative research of several ancient maps by Hapgood and his team reveals "that the land bridge was a broad one,

---

[1] Charles H. Hapgood, *Maps of the Ancient Sea Kings: Evidence of Advanced Civilization in the Ice Age*, (Adventures Unlimited Press, 2014), 40.

perhaps a thousand miles across. In case the reader is drawing back at this moment, in a state of amazement mingled with horror, I am forced to remind him that this bit of evidence is only a link in a chain. We have completed a study of the Piri Re'is Map of 1513, and have concluded that it may contain a representation of part of the Antarctic coast drawn before the present ice cap covered it. We have examined the 1531 Oronteus Finaeus Map of Antarctica and have come to much more far-reaching conclusions. We cannot estimate, of course, the lapse of time implied by these remarkable maps of Antarctica. But we have presented evidence that the deglacial or unglaciated period in the Antarctic cannot have come to an end later than 6,000 years ago."[1]

Evidence of this extremely temperate time period has also been confirmed by a recent Arctic coring expedition. Marine sedimentary sequences found in one of the drilled cores revealed sea surface temperatures near the North Pole were at some time in the past over 73 degrees Fahrenheit. Geologists credit higher-than-modern greenhouse gas concentrations along with extremely high cloud cover, as one of the main contributors to these warm temperatures. The moisture and clouds remaining after the steam explosion from the great deep would have served as a blanket for Earth, and warm conditions persisted as the world was repopulated. This temperate climate was the perfect habitat for proliferation of animals and humans as they migrated and settled

---

[1] Charles H. Hapgood, *Maps of the Ancient Sea Kings: Evidence of Advanced Civilization in the Ice Age*, (Adventures Unlimited Press, 2014), 99.

all around the globe after the confusion of language at the Tower of Babel about one hundred years post-flood.

Petroglyphs & Solar Flares

I was surprised to discover solar flares had contributed to the creation of ancient rock art around the globe. It was physicist Anthony Peratt who first suspected petroglyph carvings had been created by people witnessing a great solar outburst. To prove his point, he collected location data for several tens of thousands of places that had petroglyph drawings carved into rocks thousands of years ago. After using 3-D mapping to analyze his data, he was able to prove these patterns, carved into walls of stone around the world, were predominantly aligned in a "preferred" direction with regards to the sun. This caused him to imagine a solar outburst may have had something to do with the creation of petroglyph drawings.

He brought his hypothesis into the laboratory to find out if these patterns truly were created by people observing the impact of a high-energy plasma discharge upon their atmosphere. He discovered it had to have been a solar wind so strong it would be categorized today with a magnitude between one and two orders. He filled his laboratory with a moist, vog-filled atmosphere, then exposed these conditions to multiple high-currency, Z-pinch plasma eruptions. As Anthony Perrat simulated the effects of a huge solar flare within his laboratory, to his great surprise, petroglyph patterns were formed. One of his laboratory experiments generated patterns almost identical with the "squatter man." Peratt notes in his article published by the institute of *Electrical and Electronics Engineers*, that his laboratory had produced shapes that could only be created by a solar flare so powerful that it would "mimic closely, phenomena associated with the highest energy releases known, some of whose instability shapes were not known even a few years ago."[1]

A solar outburst this large has not been recorded since the time when these ancient petroglyphs were etched within surfaces of rock and into wooden pillars thousands of years ago. "The striking similarity of petroglyphs to plasma experiments would indicate that they are reproductions of intense electrical phenomena."[2] Peratt also points out the basic shapes created within his laboratory were recorded without any cultural bias or embellishments like those of the ancient artists, who with great wonder had etched into stone what they observed from different locations all around the globe. From the "Inga stone" in Brazil to the Easter Island "Rongorongo" tablet and all around the globe to Armenia's "Valcamonica" rock art.

---

[1] Anthony L. Peratt et al., "Characteristics for the Occurrence of a High-Current Z-Pinch Aurora as Recorded in Antiquity Part II: Directionality and Source," *IEEE* 35, no. 4 (August 2007): 778-807. https://ieeexplore.ieee.org/document/4287072.

[2] Anthony L. Peratt, "Characteristics for the Occurrence of a High-Current, Z-Pinch Aurora as Recorded in Antiquity," *IEEE* 31, no. 6 (December 2003): 1192-1212. https://plasmauniverse.info/downloads/PerattAntiquityZ.pdf.

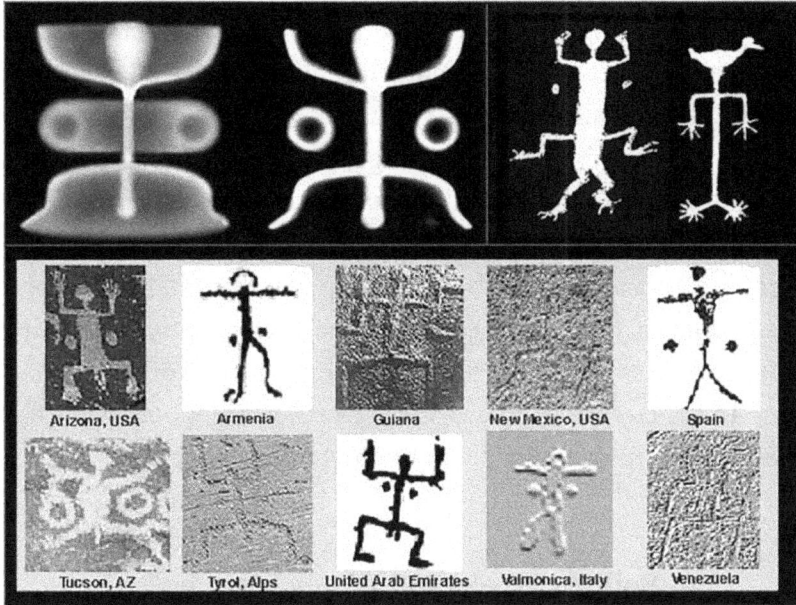

Petroglyphs Squatter Man
Credit: "Squatter man" images gathered by Anthony Peratt[1] The Squatting
Man - the Squatter Man - petroglyphs and rock art recordings of a high energy
z-pinch plasma discharge as studied by Anthony Peratt

Evidence for a massive solar storm impacting the planet at some time in the past was also found in ice cores drilled from Greenland and Antarctica. They contained traces of chemical elements indicating a massive solar storm had hit the Earth.

---

[1]http://www.thunderbolts.info/tpod/2004/arch/041231predictions
-rock-art.htm

https://www.everythingselectric.com/forum/index.php?topic=176

(this website has a color image with good resolution, also published in grey scale by Peratt in his pdf "Characteristics for the Occurrence ..." see cxi below ... not sure if ok to use permission-wise

The elements like beryllium, which increases electrical and thermal conductivity, and "chlorine" which just so happens to contain a naturally occurring radioactive cosmogenic isotope, a rare nuclide created when high-energy cosmic rays from the sun reach the Earth. "'We have studied drill cores from Greenland and Antarctica and discovered traces of a massive solar storm that hit Earth' says Raimund Muscheler, geology researcher at Lund University."[1] Ancient petroglyph inscriptions record a stunning worldwide event that had once alarmed humanity. Here is a link to Peratt's research . . . https://plasmauniverse.info/downloads/PerattAntiquityZ.pdf

A Sudden Freeze and Wooly Mammoths

The combination of volcanic outgassing and evaporating moisture from warm oceans would have created heavy vog-like conditions. Vog is created when volcanic gases and particles react with sunlight, oxygen, and moisture. Anthony Peratt's findings led me to wonder if solar outbursts had also played a role in triggering a drastic freeze, one that caused an extremely sudden "Age of Ice." A great number of woolly mammoths were frozen in an upright position. Some of them even had buttercups, leaves, and grass between their teeth.[2] A more

---

[1]Lund University, "Mysterious Solar Storm Occurred 9,200 Years Ago – Revealed by Ancient Ice," *SciTechDaily*, (July 2023). https://scitechdaily.com/mysterious-solar-storm-occurred-9200-years-ago-revealed-by-ancient-ice/.

[2]Michael Oard, *Frozen in Time*, (Green Forest: Master Books, 2004), 20.

recent publication by *LiveScience* reported researchers found a woolly mammoth so exceptionally well preserved beneath the permafrost that it oozed fresh blood as they dislodged it.[1] Not only this, but these giant creatures froze so fast that "scientists found partially preserved stomach vegetation in some of the carcasses and so could identify the woolly mammoth's last meal."[2] Evidence like this begs us to ask a couple of questions:

1) How cold does it need to be, . . . and
2) How fast does an animal the size of an elephant freeze in order for evidence of its last meal to have been preserved and reveal to us thousands of years later?

According to author Michael Oard, Birds Eye Frozen Foods Company ran calculations and came up with an answer to this question, approximately negative 150°F or -100°C.[3]

There are more than four hundred subglacial freshwater lakes discovered in Antarctica, as well as underneath Greenland's Ice Sheet and Iceland's Vatna Glacier.[4] Could it be that ice was deposited so fast it trapped these freshwater lakes beneath two miles of ice before these lakes had

---

[1] Tia Ghose, "Photos: A 40,000-year-old mammoth autopsy," *Live Science* (November 2014). https://www.livescience.com/48768-photos -mammoth-autopsy.html.

[2] Michael Oard, *Frozen in Time*, (Green Forest: Master Books, 2004), 13.

[3] Michael Oard, *Frozen in Time*, (Green Forest: Master Books, 2004), 14.

[4] "Lake Vostok mysteries: Biologists find over 3,500 life forms in isolated Antarctic basin," *RT*, (July 2013). https://rt.com/news/lake-vostok -bacteria-dna-745/.

time to freeze? It was recently discovered that water's freezing point can be changed by an electrical charge. "A study in the Feb. 5 Science reports that water can freeze at different temperatures depending on whether the surface it rests on is positively or negatively charged." The article goes on to say, "'We are very, very surprised by this result,' says study coauthor Igor Lubomirsky of the Weizmann Institute of Science in Rehovot, Israel. 'It means that by controlling surface charge, either positive or negative, you can either suppress ice formation or enhance ice formation.'"[1]

Who Survived the Ice Age

Job 38:29 says, "From whose womb has come the ice? And the hoary frost of heaven, who has given it birth?" Giving birth from the womb is a cathartic experience with unknown variables. Perhaps this God-inspired question about the womb from whence the ice came is revealing a secret glimpse into the nature of this event. Placing the Ice Age within the biblical timeline and radical post-flood climatic conditions, it seems likely this event took place sometime after the dispersion of people from the Tower of Babel. It is a great mystery, but a world map drawn in 1531 by Oronteus Finaeus coincidently associates the name of Nimrod with a glacier in

---

[1]Lisa Grossman, "Electric Charge Can Change Freezing Point of Water," *Wired*, (February 2010). https://www.wired.com/2010/02 /electric-charge-can-change-the-freezing-point-of-water/.

Antarctica. Perhaps ancient lore had something to do with naming it "The Nimrod Glacier."

The next question that came to mind was who were these people that survived a sudden freeze? Did they live in caves and as a result become known as cavemen? Were they people with wisdom, memories, and tales about their ancestors living prior to the flood? Did post-flood atmospheric conditions set the stage for a solar outburst triggering a storm of ice from the hoary frosts of heaven? After taking into consideration the sudden freeze of woolly mammoths, subglacial lakes in Antarctica with living organisms trapped beneath ice two miles thick, drilled ice cores revealing traces of a massive solar storm, ancient maps showing land bridges and the coastline of Antarctica before the ice age; placing all of this information within a biblical timeline is what led me to think that most certainly the ice age took place after the Lord had scattered His people over the face of the Earth. This was a time when a deep freeze would have forced people to seek shelter in caves and dwellings constructed of stone. Perhaps those inclined to listen for the voice of God prior to this event were led to prepare for themselves a place of shelter in advance. Amazing underground cities like Ozkonok in Turkey are believed to have been built around three thousand years ago. They were constructed to serve as emergency shelters for a few thousand people. It was in this same location archaeologists discovered two hundred more underground cities connected by tunnels. Excavations of Derinkuyu are still in progress, revealing a subterranean city eighteen stories deep with a complex system of ventilation shafts allowing air to circulate through every room. At the lowest levels there is a series of rivers and centrally located wells which provide water for the entire underground city.

Caves around the globe provided shelter for those who migrated from the Tower of Babel, with some being more elaborate than others. For example, the magnificent Longyou caves in China span over more than 300,000 square meters. Despite the harshest of conditions, people experiencing this catastrophic event left behind signs of their existence. Art drawn upon the wall of the Lascaux caves in France provides us with a unique glimpse into who these people were that took refuge in caves. Impressive drawings decorate the walls of the Sumpang Bita Cave in Sulawesi Indonesia, where human hands were airbrushed onto the walls using pigment. Photographs of this art show the detailed drawing of a pig, snake, and rabbit, adjacent to airbrushed hands on the wall of this cave. "Figurative cave paintings depicting pig hunting in the Maros-Pangkep karst in Sulawesi were estimated to be [. . .] at least 43,000 years old. The finding was noted to be 'the oldest pictorial record of storytelling and the earliest figurative artwork in the world.'"[1] Question is, were these people recording their experiences within a discernable biblical timeline of 4,300 years or does it take a leap of faith in science to imagine this took place 43,000 years ago?

---

[1]M. Aubert et al., "Earliest hunting scene in prehistoric art" *Nature* 576, (December 2019): 442-445.

Cueva de las Manos in Argentina has similar artwork with "stenciled outlines of human hands in the cave, but there are also many depictions of animals, such as guanacos (Lama guanicoe), still commonly found in the region, as well as hunting scenes that depict animals and human figures interacting in a dynamic and naturalistic manner." Cueva de las Manos "is considered to be one of the most important sites of the earliest hunter-gatherer groups in South America."[1]

According to Unesco, the Chauvet-Pont-d'Arc Cave in southeastern France is considered to be one of the most significant prehistoric art sites in the world, containing some of the best-preserved figurative cave paintings. "Hundreds of animal paintings have been catalogued, depicting at least 13 different species, including some rarely or never found in other ice age paintings."[2] I find it interesting the above quote suggests this art was drawn during the Ice Age. This cave also has stencils of hands on the walls just like the Sumpang Bita Cave. "The art is also exceptional for its time for including "scenes," e.g., animals interacting with each other; a pair of woolly rhinoceroses, for example, are seen butting horns in an apparent contest for territory or mating rights."[3] The perspective and depth of motion in these drawings reveals

---

[1] https://whc.unescoorg/en/list/936/

[2] s.v. "Chauvet Cave," *Wikipedia*, https://en.m.wikipedia.org/wiki/Chauvet_Cave.

[3] Carole Fritz and Tosello Gilles, "The Hidden Meaning of Forms: Methods of Recording Paleolithic Parietal Art," *Journal of Archaeological Method and Theory*, 14, (February 2007): 48-80. https://doi.org/10.1007/s10816-007-9027-3.

a high degree of intelligence. Can you just imagine seeing a pair of woolly rhinoceroses fighting and documenting your experience of cave life with art using fresh pigments you had cultivated from nature?

As people were forced to become hunter-gatherers, harsh conditions took a toll on those who had migrated to the far north and extreme southern regions of planet Earth. For extended periods of time these people were not exposed to sunlight, and eventually the lack of sunshine, along with poor nutrition, would have resulted in a deformation of bone structure. The pronounced brow ridge of a "caveman" may have been formed due to rickets. The medical terminology for a prominent, protruding forehead is frontal bossing. To quote from MedMD, "A condition that could lead to frontal bossing after birth is rickets. Rickets typically occur in infants and young children and are caused by a lack of vitamin D."[1] A lack of vitamin D sums up to a lack of sunshine. Professor of Natural Sciences, Robert Schoch, points out archaeological discoveries confirm there was a pause in the advancement of human culture and a noticeable decline in civilization for many years before they re-emerged in Mesopotamia, Egypt, and other areas.[2] It may well be that the advancement of humanity was paused because the great

---

[1]Dan Brennan, MD, "Front Bossing: What Is It?" *WebMD* (March 2021). https://www.webmd.com/baby/frontal-bossing-what-is-it.

[2]Robert M. Schoch, PH.D, "The Mystery of Gobekli Tepe," *Project: Yourself.*https://projectyourself.com/blogs/news/the-mystery-of-gobekli -tepe.

flood of Noah resulted in global atmospheric conditions perfect for producing an age of ice.

Years passed as rugged survivors of an ice age waited for the climate to stabilize. A dramatic rise in sea levels took place as ice packs and glaciers melted. As a result, people who had migrated great distances along the shoreline of exposed coastlines suddenly found themselves isolated on fragments of land. As they ventured from their familiar cave dwellings to explore the world once again, it was a whole new experience. After the ice had sufficiently melted, Earth was burgeoning with abundant life. But a question remains, do we have concrete evidence for human history stretching back 43,000 years, or is it more likely our ancestors survived a post-flood ice age taking place about 4,300 years ago according to biblical chronology? Ages described in the millions and billions of years are not within the scope of reliable observation, and it seems to me many conclusions are based upon mere presumptions that Earth is extremely old.

CHAPTER 11

SHATTERING
SIGNIFICANCE

As people adapted to their new surroundings unique talents and cultures emerged. The discovery of elaborate beads made of glass, gold, silver, and semiprecious stones reveal a sophisticated knowledge of pyro technology existed thousands of years prior to Christ's birth.

> *One of the earliest and most startling discoveries of the archaeologists was a lump of manufactured opaque blue glass at Eridu, sixteen kilometers southeast of Ur. The glass had been produced well before 2000 B.C.E., a dating which was fixed by the fact that the glass had been found underneath a pavement that had been laid down during the reign of Aram-Sin, the third king of the third dynasty of Ur. Even more ancient, manufactured glass artifacts were subsequently unEarthed from Ur itself [. . .] The existence of glass artifacts manufactured at such an early time brought the scientific community face to face with a fact of shattering significance [. . .] glass*

CHAPTER 11: SHATTERING SIGNIFICANCE

manufacture requires a pyro technology that was presumably yet a thousand years away, an Iron Age technology that had never been considered to have originated in Mesopotamia.[1]

Ebla was one of the earliest kingdoms in northern Syria occupied over five thousand years ago. Thousands of cuneiform tablets were discovered in 1974, and scholars say they date back to a time prior to the Akkadian Empire. These cuneiform archives of Ebla record transactions and trade agreements with the surrounding locations. They also contain names of biblical figures and, as a result, are very useful for our understanding of this time period. The discovery of these archives of Ebla confirms the accuracy of Holy Scripture. Ancient Ebla was a literate, organized, and powerful empire. After digging down fifty feet in some places archaeologist Leonard Woolley points out that "the tomb objects themselves show that art had already reached a level which would have been impossible without a previous history. [. . .] On the technical side alone, the knowledge of metallurgy proved by the use of alloys and the skill shown in the casting of these alloys is remarkable and was assuredly not acquired in the course of two or three generations."[2] The First Dynasty of Ur was "supposedly" the third one after the flood. It was at this location where archeologists and interpreters Giovanni Pettinato and Paolo Matthiae

---

[1]Samuel Kuninsky, *The Eighth Day*, (Jason Aronson, Inc., 1994), 32.

[2]C. Leonard Woolley, *The Sumerians*, (New York and London: Oxford University Press, 1965), 42.

were met with opposition. There were great discussions and disagreements between scholars regarding a possible connection between the Syrian city of Ebla and the Bible. It sounds to me like Pettinato and Matthiae were lassoed into a spiritual battle for the truth. The discovery of three tablets inscribed with almost the same text were deciphered by the interpreter, Giovanni Pettinato, who comprehended their inscriptions to be an Eblaite creation hymn: "Lord of heaven and Earth: The Earth was not, you created it, The light of day was not, you created it, The morning light you had not [yet] made exist."[1]

A similar discovery to the one made at Ebla was in Mari, about two hundred miles southeast of Haran. Richie Cooley asserts in his book named *Is the Bible Divinely Inspired (Special Edition)* that "Mari shares a common culture with the area where the Patriarchs originated" (pg. 40-41). Both Ebla and Mari controlled large territorial empires adjoining each other in the middle of the Euphrates.[2] Archaeological research reveals this great city of Mari was built to control the waterways of the Euphrates connecting trade routes between the Eastern Mediterranean and Southern Mesopotamia. Mari was an ancient Semitic city in Syria that flourished between 2900 BC and 1759 BC—Keller goes on to say that "Assyriologists dealing with these reports of governors and district commissioners of the Mari empire came

---

[1] s.v. "Ebla—biblical controversy," *Wikipedia*. https://en.m.wikipedia.org/wiki/Ebla-Biblical_controversy.

[2] Cyrus Herzl Gordon (Editor), *Eblatica*, (Eisenbrauns, 1992), 69.

across one after another a whole series of familiar sounding names from the biblical history—names like Peleg, and Serug, Nahor and Terah, and—Haran." Over seventeen thousand cuneiform tablets discovered at the ancient site of Ebla, reveal a wealth of detailed information about commercial and political relations with the kingdoms of Sumer, Kish, Ur, and Egypt between 2500 and 2300 BC. According to biblical chronology, this was some time after dispersion of people from the Tower of Babel. Werner Keller says in his book *The Bible is History*, "Scholars had for a long time been familiar with the royal city of Mari which features in many old inscriptions from Babylonia and Assyria. One text maintained that Mari was the tenth city to be founded after the flood" (pg. 61).

Mari is located about 200 miles (320 km) southeast of Haran. The Mari tablets also describe the city of Haran as a flourishing community ideally located as a hub for important trade routes running from east to west. The city of Tehran, named after Abraham's father Terah, is currently a thriving metropolis. Ur, the land where Job dwelt, is known today as Uzbekistan. The descendants of Noah's son Shem later became known as the Semites ("Sem" is the Greco-Latin form of Shem). Shem's sons, Elam and Asshur, became known as the Elamites and Assyrians (a.k.a. Assurians). Texts from the ancient city of Ebla and Mari also record the names of Noah, Abram, Laban, and Jacob.

The Weld-Blundell Prism is a clay, cuneiform, inscribed vertical prism containing the Sumerian King List. "The prism was found in a 1922 expedition in Larsa in modern-day

Iraq by British archaeologist Herbert Weld Blundell."[1] This amazing artifact not only records information about rulers of Mesopotamia prior to 1800 BC, it also has stories about those who ruled prior to the great flood of Noah. A time when people lived to be eight or nine hundred years old. The list is very complex, but when the names are carefully read, it becomes apparent some of the rulers were contemporary and governed alongside each other. It is believed the flood event, as recorded by the King List, notes a separation between the historic, genealogical, timeline recorded by Holy Scripture, and a mythical time when legends of dragons and fairies were implemented to lull humanity into a great deception. There are several versions of the Sumerian King List with content variations about the names of kings or lengths of their reigns, and it is believed that "these differences are both the result of copying errors, and of deliberate editorial decisions to change the text to fit current needs."[2] So apparently some of the changes were made to suit the whims of whomever ruled at the time. Most scholars label this time prior to the flood as mythical, but solid evidence for people, places, and events keep bubbling to the surface with archaeological findings made known to the world by the "information age."

---

[1] s.v. "Weld-Blundell Prism," *Wikipedia*, https://en.m.wikipedia.org /wiki/Weld-Blundell_Prism.

[2] s.v. "Weld-Blundell Prism," *Wikipedia*, https://en.m.wikipedia.org /wiki/Weld-Blundell_Prism.

Abraham

Genesis 11:31 records Abraham was from Ur Kasdim, which is also referred to as "Ur of the Chaldees" in Genesis 15:7. Historic records reveal the "Chaldees" were a nomadic tribe who did not rule Southern Mesopotamia until 1,500 years after Abraham's time, around 600 BC. In *The Bible Knowledge Commentary: Old Testament,* it is brought to our attention that "the Sabeans and Chaldeans (Job 1:15, 17) were nomads in Abraham's time, but in later years they were not nomadic."[1] Genesis chapter 14:13 unequivocally categorizes Abraham as a "Hebrew," and not as a "Chaldean." The term Ur Kasdim, when applied to Abraham's time period, more accurately describes the surrounding influences of Abraham's upbringing. The book of Jasher delves into greater detail than the Bible *about the surroundings of Abraham as a child, and describes his father, Terah, as the son of Nahor, and he was "a prince of Nimrod's host." Terah was "very great in the sight of the king and his subjects, and the king and princes loved him, and they elevated him very high" (The Book of Jasher 7:49). Nimrod dignified Terah and elevated* him above all his other princes (Jasher 7:41).

The Hebrew term "Kasdim" describes a priestly class of "Astrologers" who depended on revelation through the stars. These were the sorcerers and magicians of the day, like those who were called upon by Pharaoh when his spirit was troubled in the morning, "so he sent and called for all

---

[1] John F. Walvoord and Roy B. Zuck, *The Bible Knowledge Commentary,* (SP Publications, Inc., 1985), 717.

the magicians of Egypt" (Genesis 41:8). There are many places in Scripture where the term "Kasdim" is used. Like in Exodus 7:11 when the Pharaoh calls for his wise men, sorcerers, and magicians. Then again when frogs covered the land of Egypt. This was an extremely disturbing experience, but instead of turning to God, the Pharaoh chose to call upon his magicians with their secret arts (Exodus 8:7). Again, after Nebuchadnezzar besieged Jerusalem, he chose Daniel (Belteshazzar), Hananiah (Shadrach), Mishael (Meshach), and Azariah (Abed-nego) to be his personal servants because "for every matter of wisdom and understanding about which the king consulted them, he found them ten times better than all the magicians and conjurers who were in all his realm" (Daniel 1:20). Although Abraham was keenly aware of and influenced to believe the stars were for signs and wonders, he chose to consult the Creator of them.

Greek translators of the Old Testament thought of Mesopotamia as the homeland of Abraham around the ancient city of Haran. In Genesis 24:4 Abraham sends his servant to the city of Nahor in northwestern Mesopotamia, calling it "my country" and "the land of my birth." Cuneiform tablets from Ugarit, Nuzi, and Ebla record dealings with Ur in Northern Mesopotamia. The tablets refer to Ur, URA, and Urau as cities close by, and they were given the names of Abraham's brothers, Haran and Nahor. Wikipedia states, "according to Jewish and Muslim tradition, Urfa is Ur Kasdim, the hometown of Abraham."[1] Haran is located

---

[1] s.v. "Urfa," *Wikipedia*. https://en.wikipedia.org/wiki/sanliurfa.

ten miles north of the Syrian border in Turkey along the Balikh River, a tributary of the Euphrates. It was an important Hurrian center, where the moon god named "Sin" was worshiped. Two cities not far from Haran are Urfa and Ura. According to local tradition, Abraham was born in Urfa, which is known today as Sanliurfa in southern Turkey.

Josephus refers to Abraham as being very wise and quotes Berosus as saying, *"In the tenth generation after the flood, there was among the chaldeans a man righteous and great, and skillful in the celestial science."*[1] Abraham's birthdate is calculated to be somewhere between 2166 and 1948 BC. If the great flood ended 2345 BC and Noah lived another 350 years after the flood, Noah was still alive during Abraham's lifetime. The Book of Jasher 9:6 records that "Abram was in Noah's house thirty-nine years, and Abram knew the Lord from three years old, and he went in the ways of the Lord until the day of his death, as Noah and his son Shem had taught him." This means Abraham would have received first-hand knowledge about everything that happened from the first day of creation. Noah corresponded with Enoch's son Methuselah for hundreds of years, who in turn had known Adam for hundreds of years. Noah's son Shem lived for 502 years after the flood and was alive when Abraham became the father of his first-born son, Isaac. Abram became known as Abraham when God promised he would be a father to the nations of our world (Genesis 17:5). His father, Terah, was an idol worshiper.

---

[1]Josephus Flavius, *Antiquities of the Jews*, (Blacksburg: Unabridged Books, 2011), 1.158.

Growing up Abram had a keen understanding of astronomy, but it was within his heart to know the Creator of the sun, the moon, and stars. Abram understood God Almighty, He alone, was worthy of worship and praise (Jasher 9:11-19).

After his father's death in Haran, Abram received God's promise of a great inheritance and set out for Canaan (Genesis 12:1-4). This was about four hundred years after the flood, and biblical records reveal that during this time, giants once again roamed the Earth. The Bible names over thirty different tribes of giants: Amalakites, Amorites, Anakims, Ashdothites, Avvims, Canaanites, Caphtorim, Ekronites, Emims, Eshkalonites, Gazathites, Geshurites, Gibeonites, Giblites, Girgashites, Gittites, Hittites, Hivites, Horims, Horites, Jebusites, Kadmonites, Kenites, Kenizzites, Maachathites, Manassites, Nephilim, Perizzites, Philistines, Rephaims, Sidonians, Zamaummins, Zebusites, and Zuzims.

The Amorite giants are mentioned more than eighty times in Scripture, and Genesis 14:13 tells us they were allies with Abram when his nephew Lot needed to be rescued. "Then a fugitive came and told Abram the Hebrew. Now he was living by the oaks of Mamre the Amorite, brother of Eschcol and brother of Aner, and these were allies of Abram." The Amorites were descendants of Noah's grandson Canaan through the lineage of Ham (Genesis 10:15-16). The Bible does not give details as to how or why this happened, so we can only postulate that the genetic influence of fallen angels was carried forward through one of the wives of Noah's sons. The Book of Amos gives us a description of the Amorite giants as the Lord spoke:

Yet it was I who destroyed the Amorite before them, whose height was like the height of the cedars, and he was as strong as the oaks; yet I destroyed his fruit above and his roots beneath. Also, it was I who brought you up from the land of Egypt, and led you forty years through the wilderness, to possess the land of the Amorite (Amos 2:9-10).[1]

Fragment of a wall painting from the palace of Mari, "Organizer of Sacrifices." The leading figure is at least twice the size of those following.

After God revealed the Promised Land to Abram, and he saw all this land stretched out before him (Genesis 13:15), why did he choose to dwell by the oaks of Mamre? This is where

---

[1]https://commons.wikimedia.org/wiki/File:L%27Ordonnateur_du _sacrifice_-_Mus%C3%A9e_du_Louvre_Antiquit%C3%A9s_orientales _AO_19825.jpg

the giant Amorites were! "And there he built an altar to the LORD" (Genesis 13:18). This was not just a bold demonstration of Abram's faith but it was also symbol of God's faithfulness to keep His promise. For it was here that God not only made a covenant with Abraham but a promise to all humankind that by His power, the giants would be overcome. It was a covenant based upon God's faithfulness and not dependent upon human efforts. For just as Abraham believed God and it was credited to him as righteousness, we too should recognize our faith makes us sons of Abraham, and like him we will be BLESSED. As God did with Abraham, by faith we can rest assured the giants in our life will be overcome or made to be allies for "Abraham BELIEVED GOD, AND IT WAS RECKONED TO HIM AS RIGHTEOUSNESS. Therefore, be sure that it is those who are of faith who are sons of Abraham. And the Scripture, foreseeing that God would justify the Gentiles by faith, preached the gospel beforehand to Abraham, saying, 'ALL THE NATIONS SHALL BE BLESSED IN YOU'" (Galatians 3:6-8).

Genesis 15:12-21 records terror and great darkness fell upon Abraham as he heard God saying there would be oppression and slavery, and the iniquity of the Amorite giants was not complete. Then God seals the covenant before naming nine other tribes of giants. These prophetic verses of Scripture were encouraging Abraham to have faith in the face of adversity, as well as unveiling a revelation about what would take place in the future. The Abrahamic Covenant establishes that every person is saved by grace through faith.

SUMMARY

In the Old Testament faith was expressed by believing in God. With the New Testament our faith is in Jesus because He fulfilled all of the Old Testament prophecies about a sacrificial lamb. Jesus is the Lamb of God (John 1:36), who was slain to pay the price for our sins and now sits upon the throne at the right hand of God the Father (1 Peter 3:22). Scriptures say we will be raised up just as Christ was risen from the dead. "Now God has not only raised the Lord, but will also raise us up through His power" (1 Cor 6:14). The power God gives us is according to our faith, and now we as "believers" are to live "in confident expectation of things yet to come, to walk by faith and not by sight" (2 Cor. 5:7). The key words in this verse seem to be "confident expectation" for we are to have complete confidence in God's promise, even as Adam did after he fell from grace.

The information age is most certainly here as an invisible stream flows beneath the surface of our consciousness with torrents of data and ideas sweeping us to and fro. Like a fire hose with the pressure of a machine gun forcefully slamming loose branches of a giant tree against the house, information is shooting at us now with alarming force and tenacity. Question is . . . how do we dodge the bullets of misinformation? Who are you going to believe? What sources of information will you follow? Upon what will you focus your attention and base what may turn out to be life-transforming decisions? As a society, we have not been encouraged to question the information presented to us by so-called "experts," nor have we been well equipped with skills to reason about or question what they have to say. Especially when it comes to the realm of science. As a result, we are led to believe by faith in scientific theories. Many of these theories are based upon concepts sprouting from circular reasoning.

I had no idea what circular reasoning was or what questions to ask until I was inspired to begin research for this book in 1993. What I discovered was scientific theories, like the layers of an onion, were building one idea upon another in order to substantiate concepts that originated without solid, observable evidence. Like radiocarbon dating, for example. When applying radioactive decay or carbon-14 dating to non-living things such as fossils, scientists base their calculations upon assumptions about a "model age," which "requires that the initial number of daughter atoms (Do) be known. No analytical equipment can give this value."[1] So because they have no idea what the initial number of atoms is, they use a location where the fossil is found within strata layers to come up with a date. Can you see how this theory is built upon another?

---

[1]Steven A. Austin, *Grand Canyon: Monument to Catastrophe*, (Institute for Creation Research, 1989), 119.

Question is, has a corruption of knowledge paved the way for a great deception? "For there shall arise false Christs, and false prophets, and shall shew great signs and wonders; insomuch that, if it were possible, they shall deceive the very elect" (Matthew 24:24). I encourage you to take pause and genuinely consider a new way of perceiving creation. The universe we observe is intelligible to us because we were created in God's image and likeness.

So does it really matter that we know where we came from? I believe it does because an accurate understanding of history cultivates discernment, nurtures our common roots, and establishes a culture where compassion, kindness, and charity prevail. What if we truly are the surviving brothers and sisters of a noble human family? It is essential to become aware of how we've been taught to think and start questioning a corruption of knowledge with regards to our ancestry, our history, and our true identity. Since the beginning of time we have been influenced by a matrix of deceit and corruption intended for the sole purpose of dividing, conquering, and gathering the spoils. We are relentlessly misled to believe that ancient monuments, artifacts, carved tablets, and historic records of Holy Scripture are myth. Our solidarity is undermined as we are robbed of our history having deep roots as a family created in the image and likeness of God. We have an inheritance to claim, the promised gift of eternal life and the restoration of our relationship with our Heavenly Father.

These are perilous times of precipitous change, and the father of lies has managed to conceal the truth about our origin and human history. His manipulation of time has played a huge role in undermining people's confidence in the authenticity and accuracy of Holy Scripture. But what if you happened upon a thread of truth woven throughout the myths and legends of old, and as you pulled upon this thread, it began unraveling a fabric of lies? As you continued to pull, an entire cloak of deception began unraveling thread by thread until the whole garment fell apart. This was my experience and why I am passionate about sharing my discovery with you now.

The biblical chronology of people and events from the time of creation to our present day adds up to about six thousand years. Today we not only have amazing archeological discoveries but Earth's geology, physics, chemistry, and even maps drawn by ancient cartographers have endured the ravages of time, proving the legitimacy and accuracy of Holy Scripture. Many discoveries from ancient times have been hidden for thousands of years but somehow were preserved, have remained intact, and are currently available for our viewing and discretion. As a result, many "old-school" theories can no longer be accepted as fact and without question. The concept of humans evolving from apes is now being challenged by a flood of evidence revealing very sophisticated, intelligent civilizations navigated the Earth thousands upon thousands of years ago. As archaeologist Paolo Matthiae says, "The sort of prehistoric anthropology into which the archaeology of the ancient Near East had so often lapsed had now been rejected, since the cultures concerned were now fully historical."[1]

---

[1] Paolo Matthiae, *Ebla: An Empire Rediscovered*, (Doubleday & Company, Inc., 1977).

What he is saying is that recent discoveries prove, beyond the shadow of a doubt, that ancient humans were not prehistoric by any means or sensibility because "observable" evidence presents a completely different story about who they were.

I pray for a foundation of wisdom to be restored, not only by His word, but with keen observation of His works. By studying God's creation, we are led to know Him. From the swirling Milky Way to rolling waves of the ocean, from the spiraling patterns of seeds in a sunflower to the twisting strands of DNA, the language of God is spoken to us through precise, mathematical patterns and equations like the golden ratio, the cosmic constant represented by the Greek symbol pi. Studying the works of His creation is instrumental for developing wisdom. "Majestic are the works of the LORD, those who delight in them, study them" (Psalm 111:2).

Many innovations and discoveries were inspired by close observation of God's creation. One recently made discovery resulted in the creation of Velcro. After walking in the woods, the Swiss outdoorsman named George De Mestral was curious to understand how cockleburs clung to his clothing so tightly. After close investigation of the lowly burr, he was inspired to create a product we all know and use today. As of 2022 the global Velcro market size was estimated to be worth more than 2,400 million US dollars.

Sir Francis Bacon gave credence to the scientific method with his assertion that all hypotheses and experiments need to be established upon "observable" evidence rather than logic-based arguments. He once said no person can be too knowledgeable about God's word or too well-studied in the book of God's works "to conclude, therefore, let no man upon a weak conceit of sobriety or an ill-applied moderation think or maintain that a man can search too far, or be too well studied in the book of God's word, or the book of God's works, divinity or philosophy; but rather let men endeavor an endless progress or proficience in both."[1]

So where does our disbelief come from? Are we writing off the significance of historic records of Holy Scripture because we have been indoctrinated to believe ideas based upon logic rather than observation? "In stark contrast to deductive reasoning, which had dominated science since the days of Aristotle, Bacon introduced inductive methodology—testing and refining Hypotheses by observing, measuring, and experimenting."[2] Theories of evolution and prehistory are just that, theories. "So what did the evolutionists do? Any history text that referred to the flood or creation less than 10,000 years ago is disregarded as myth. As a result all the ancient history books have been discarded. They have not been destroyed, just covered up. We now have a new period called 'pre-history.' There really is no pre-history, only history that they do not want us to know about."[3] Paul tells us in Romans chapter one that since the beginning of time God has made known His invisible qualities, His divine nature . . .

---

[1] Michael Kiernan, *The Oxford Francis Bacon IV: The Advancement of Learning*, (Oxford University Press, 2000).

[2] "Francis Bacon, 1561-1626," https://lib-dbserver.princeton.edu /visual_materials/maps/websites/thematic-maps/bacon/bacon.html.

[3] Ken Johnson, *Ancient Post-Flood History*, (2010), 6.

for all to see and observe through His creation. So we are without excuse; there is no denying we were created in the image and likeness of a great and mighty Creator.

With the eyes of a child I contemplated the wonder of God's creation while observing stars as they hung suspended upon a dark blanket of the night sky. I was also amazed to watch seeds sprouting from under the ground. They began to sprawl and twist into vines, producing crunchy, edible pods of sweet peas. These experiences as a child triggered my curiosity about God. As I matured and became an adult, my concept of God became more symbolic and mythological. The recorded stories of Holy Scripture appeared to be more of a fairytale than fact. It seemed ridiculous to think people lived to be eight or nine hundred years old. Now, I not only believe it was possible for people to have lived this long, I understand how it was possible with the addition of barometric pressure exerted by the vaulted dome. Today I am so grateful to be convinced about the literal accuracy of Holy Scripture and cannot help myself from questioning how people lost their sense of wonder about God? I think the answer is a great deception has been and is being played. Now is the time for logic, critical thinking, and comprehension of God's Word to be considered as profoundly accurate and true. "My people are destroyed for a lack of knowledge" (Hosea 4:6).

The time we are living demands a people with vision, those who have come to terms with the fact we are fighting a spiritual battle. For any of us to have enough courage to open our eyes and see what is coming will require complete confidence in Christ's victory over evil. A magnificent delusion is on the horizon. One that captivates people with grand assurance, and while people are saying "Peace and Safety," destruction will come upon them suddenly, as labor pains on a pregnant woman, and they will not escape (1 Thessalonians 5:3). This is the beginning of the time of trouble, but rest assured, "for God has not destined us for wrath, but for obtaining salvation through our Lord Jesus Christ" (1 Thessalonians 5:9). Perhaps our travail in this human form serves a specific purpose, one of becoming acquainted with Christ's suffering so we could possibly know how high a price was paid for our sin. Hopefully, whatever distressful challenges we go through in this life, they will lead us to a deeper understanding of the sacrifice He made for the redemption of our souls. We should never take His gift lightly. Almighty God is presenting us with an opportunity to experience the weight and substance of His eternal glory. "Those who look to Him are radiant—their faces are never covered with shame" (Psalms 34:5).

God's heart for us as children created in His image and likeness is revealed to us through His message and plan of redemption. His promise to Adam and Eve for deliverance from sin is one that has prevailed throughout all of human history. Just as God provided a way of escape for Noah and his family, He is preparing a way of escape for His bride. We know and have complete confidence in God's promise. For His Son is going to return and rapture His bride. It is prophesied that in the last days, there will be those who humble themselves in the sight of the Lord and accept Jesus as their savior. At the last minute there will be those who finally realize how great a sacrifice was made to pay the price for their sin, and at the heart of their transformation God's Will takes dominion and is manifest here on Earth as it is in heaven through those who believe and trust in Him. "Thy will be done here on Earth as it is in heaven" (Matthew 6:10). His will is being done through people who understand we were created in His image. Through those who sincerely believe and desire to have a relationship with their Creator, "behold, I stand at the door and knock" (Revelation 3:20).

The door He knocks upon just so happens to be the door of your heart. Believing and trusting in His gift of redemption prepares your heart and mind for the power of His Holy Spirit to reign in you. And as we are restored through mercy and grace, His Holy Spirit empowers us to live, and move, and have our being. To be Holy as He is Holy for His Name is above all Names. Come quickly, is my prayer to the Great and Mighty Redeemer of lost souls . . . our Heavenly Father who sacrificed His only begotten Son. I pray for an end to all suffering and for this seminal work to inspire those wiser than I to dive deeper into the conceptual framework presented here.

Now is the time to become spotless, blameless, and made pure as we anticipate His return. Persevere, remain strong and powerful by establishing an intimate relationship with God through His only begotten Son, who is also known as Emmanuel, meaning "God With Us." "For God hath not given us the spirit of fear; but of power, and of love, and of a sound mind" (2 Timothy 1:7). Through prayer and the contemplation of His Word we can build a relationship with him. This needs to be our number one priority as we are His bride awaiting with great expectation for His return with complete confidence in and assurance of His promise. We know He transcended this physical dimension and went to prepare a place for us. "Do not let your hearts be troubled. You believe in God; believe also in me. In My Father's house are many rooms. If it were not so, would I have told you that I am going there to prepare a place for you?" (John 14:1-2). Therefore, seek His face, expect His return, and be filled with the Joy of the Lord. Those who accept Jesus paid the price for their sin will not suffer tribulation but will be transformed in the twinkling of an eye.

> Listen, I tell you a mystery: We will not all sleep, but we will all be changed—in a flash, in the twinkling of an eye, at the last trumpet. For the trumpet will sound, the dead will be raised imperishable, and we will be changed. For the perishable must clothe itself with the imperishable, and the mortal with immortality" (1 Cor 15:51-53).

We know, beyond the shadow of a doubt, the Spirit of the Living God will be with us to the very end. God did not appoint His beloved bride to suffer wrath. Those who choose to believe in Him will be caught up when "the Lord Himself will descend from heaven with a shout, with the voice of the archangel, and with the trumpet of God [. . .] then we who are alive and remain shall be caught up together with them in the clouds to meet the Lord in the air, and thus we shall always be with the Lord. Therefore, comfort one another with these words" (1 Thessalonians 4:16-18). I plead with you to fix your eyes upon Jesus. Do not reject the Father of your salvation from the beginning of creation for all have sinned and none of us is better than another (Romans 3:19). He is the author and perfecter of our faith, "who for the joy set before him endured the cross, scorning its shame, and was seated at the right hand of the throne of God" (Hebrews 12:2).

A great confusion has been promoted by mainstream media with the concept of an apocalypse. This word has been twisted by Hollywood to represent doom and gloom, but the actual meaning of it is to uncover, to disclose or unveil. It is the unveiling of a great mystery. The book of Revelation is the one and only book of the Bible with a promised blessing for those who read and understand it. Revelation unveils the power of Christ Jesus returning for His bride who endures with loving expectation of His appearance. Those who refuse Jesus as the way, the truth, and the life (John 14:6) will find themselves in the middle of a great tribulation. When they reject the Truth, and choose to not believe and accept the Lamb of God as their Lord and Savior, they most certainly will endure suffering before realizing they have been deceived. Just as John the Baptist records in the book of Revelation 7:14-17.

And he causes all, the small and the great, and the rich and the poor, and the free men and the slaves, to be given a mark on their right hand or on their forehead, and he provides that no one should be able to buy or to sell, except the one who has the mark, either the name of the beast or the number of his name" (Revelation 13:16-17).

Programmable digital currency implemented and governed by the beast will make it impossible to buy or sell without his mark. Do not be ignorant about the ways and wiles of Lucifer whose agenda is to deceive humanity with grand delusions. As Matthew 24:24 records—even the very elect will be deceived. "While they are saying, 'Peace and safety!' then destruction will come upon them suddenly" (1 Thessalonians 5:3). This great deception portends all nations will be fooled by a false messiah as he is seated upon the throne of the temple in the holy of holies.

The GIFT of salvation awaits you, and as Pastor JD Farag of Kaneohe Chapel explains, it is easy to receive this gift. It is simple as ABC . . .

A. Admit you have sinned and know that you are not alone in this admission, for all have sinned (Romans 3:23).
B. Believe Jesus died on the cross to pay the price for your sin, and . . .
C. Call upon His name. Trust in the Lord with all your heart and He will give you rest, for it is anxiety that leads us into temptation and disobedience.

The End

www.ingramcontent.com/pod-product-compliance
Lightning Source LLC
Chambersburg PA
CBHW060049100426
42742CB00014B/2746